Withdrawn

Other titles in this series

The Moon and How to Observe It
Peter Grego

Double & Multiple Stars, and How to Observe Them
James Mullaney

Saturn and How to Observe it
Julius Benton

Jupiter and How to Observe it
John McAnally

Star Clusters and How to Observe Them
Mark Allison

Nebulae and How to Observe Them
Steven Coe

Galaxies and How to Observe Them
Wolfgang Steinicke and Richard Jakiel

Related titles

Field Guide to the Deep Sky Objects
Mike Inglis

Deep Sky Observing
Steven R. Coe

The Deep-Sky Observer's Year
Grant Privett and Paul Parsons

The Practical Astronomer's Deep-Sky Companion
Jess K. Gilmour

Observing the Caldwell Objects
David Ratledge

Choosing and Using a Schmidt-Cassegrain Telescope
Rod Mollise

Martin Mobberley

Supernovae
and How to
Observe Them

with 167 Illustrations

 Springer

Martin Mobberley
martin.mobberley@btinternet.com

Library of Congress Control Number: 2006928727

ISBN-10: 0-387-35257-0   e-ISBN-10: 0-387-46269-4
ISBN-13: 978-0387-35257-2   e-ISBN-13: 978-0387-46269-1

Printed on acid-free paper.

9 8 7 6 5 4 3 2 1

springer.com

To Tom Boles, Mark Armstrong, and Ron Arbour who, together, have discovered almost 200 supernovae from the very cloudy skies of the U.K. A remarkable achievement!

Preface

Supernova explosions, which mark the deaths of massive stars or of white dwarf stars in binary systems, are unbelievably violent events. Despite occurring in galaxies many millions of light-years away, amateur telescopes can reveal these colossal explosions, and even discover them. In the past 25 years, the amateur astronomer's contribution to supernova research has been staggering. Visual variable star observers with access to large-aperture amateur telescopes have contributed a steady stream of magnitude estimates of the brightest and closest supernovae. In addition, with the increasing availability of robotic telescopes and CCD technology, more and more amateurs are discovering supernovae from their backyards. Worldwide, there have been more than 400 supernovae discovered by amateur astronomers using amateur telescopes. Supernova research has never been so important. Recent professional measurements of the most distant Type Ia supernovae have revealed the staggering and unexpected discovery that the acceleration of the Universe is actually increasing! This, in turn, has led to a new phrase, *dark energy*, entering the astronomical vocabulary; a mysterious force, in opposition to gravity, driving the accelerated expansion. Although amateurs cannot study the farthest supernovae, their discovery and measurement of the closer examples helps to refine the science that is the hottest topic in cosmology today; that is, pinning down the history of the Universe and how much mass and energy exists within it today. As always, amateur astronomers are making a valuable contribution, and, hopefully, this book might inspire a few more to monitor and discover new supernovae.

Martin Mobberley
Suffolk, U.K.
October 2006

Acknowledgments

As was the case with my previous three Springer books, I am indebted to the outstanding amateur astronomers who have donated images and advice to this new work. In alphabetical surname order, the help of the following supernova experts, amateur astronomers, and observatory/science facility staff is gratefully acknowledged: Ron Arbour; Mark Armstrong; Adam Block; Tom Boles; Kathie Coil; Allan Cook; Jamie Cooper; Ray Emery; Bob Evans; Alex Fillipenko; John Fletcher; Gordon Garradd; Maurice Gavin; Robert Gendler; Sergio Gonzales; Dr. Mario Hamuy; Arne Henden (AAVSO); Guy Hurst; Weidong Li; David Malin; Brian Manning; Berto Monard; Stewart Moore; Bill Patterson; Saul Perlmutter; Gary Poyner; Tim Puckett; Gordon Rogers; Michael Schwartz; Ian Sharp; Jeremy Shears; Daniel Verschatse.

I am also indebted, in various ways to the following organizations: *The Astronomer* magazine; The British Astronomical Association (BAA); NASA; ESA; STSCI; AAO; Carnegie Supernova Project; NOAO; CBAT.

I am also indebted to my father, Denys Mobberley, for his tireless help in all my observing projects. Sadly, my mother Barbara died of bowel cancer on September 8th 2006. Her support, while I was writing this book, and in all my astronomical endeavours, was always 100%.

Finally, many thanks to Jenny Wolkowicki, Harry Blom, Chris Coughlin and John Watson at Springer for making this book possible. Without Jenny's expertise and attention to detail I dare not imagine what state the final book would have ended up in!

Contents

Part I

Supernovae: Physics and Statistics

Chapter 1

Supernova Physics

Introduction

Even without a working knowledge of astrophysics, the term *supernova* conjures up a vision of an almighty stellar explosion, even amongst non-astronomers. The term was first used by Fritz Zwicky (1898–1974) and Walter Baade (1893–1960), two pioneers of the photographic era. Zwicky himself was, by all accounts, a somewhat abrasive character who once stated that the other astronomers at the Mount Wilson Observatory were "spherical bastards." When asked to explain the use of the word "spherical," he allegedly explained that they were bastards when looked at from any angle! Abrasive or not, Zwicky was the first obsessive supernova hunter, and his vision of these events being almighty stellar explosions was accurate; indeed, an explosion on the scale of a supernova is truly beyond our capacity to comprehend. All one can do is juggle with huge numbers, containing endless zeroes, and pretend we understand the scale of events involved. But before we look at what a supernova really is, let us explain a few basic concepts here so that readers who are relatively new to astronomy will not get lost.

Supernovae are stars that have reached the end of their life in a very dramatic fashion, but they are, essentially, just stars like our own sun. Okay, many potential supernovae are actually half of a binary star system and many would make our sun look very small indeed, but they are stars just the same. As our sun is not in a binary system and not a massive star, it will end its life far more peacefully. Stars exist in and around galaxies like our own Milky Way, where there is enough matter to form objects that big. Our own galaxy is approximately 100,000 light-years across (see Figure 1.1) and contains more than 100 billion stars. We are only 4.2 light-years from the nearest star, but our Milky Way Galaxy is more than 2 million light-years from the nearest big galaxy (Andromeda, or M 31). Supernovae are both rare and common events. This does sound highly contradictory, but please read on! They are rare because even stars that are destined to become supernovae may last billions of years before the final day comes, yet the flaring up and dying down of the star will last mere months. So how can they be common, too? Well, roughly 300 supernovae are discovered every year because professional astronomers and advanced amateur astronomers patrol 10,000 to 20,000 galaxies on a regular basis. With each galaxy containing a hundred billion stars, the chances of discovering a supernova rapidly improve when you scour thousands of them every clear night.

Mentally grasping the size of the visible Universe is virtually impossible. Note I say *visible* Universe. We can only see the objects whose light has had time to reach us in the 14 billion years since the Universe formed. But, to make things simple, let

Figure 1.1. Our own galaxy is thought to be roughly 100,000 light-years across, but recent infrared images from NASA's Spitzer Space Telescope indicate that it may have a bar, 27,000 light-years long at the center. Image: Spitzer Science Centre/JPL-Caltech/NASA.

us just imagine that the most distant objects we can see are 10 billion light-years away. If we pretend that this is the same as 10,000 km on the earth's surface, then a light-year becomes a millimeter and the nearest star to us is 4 mm away. Our galaxy then becomes a disk 100 m across, and the nearby Andromeda Galaxy is 2 km away. In practice, galaxies that can be patrolled by amateurs then lie within 1,000 km of us, and we are looking for the explosion of an object well under a thousandth of a millimeter across (even for a massive star). It makes you think, doesn't it!

Modern computer-controlled telescopes, both amateur and professional, enable hundreds or thousands of galaxies to be imaged each night. This nets a new supernova every day or so, on average. With such a huge number of supernovae having now been found (almost 4,000 at the time of writing), we have a huge amount of information about them. Therefore, their basic structure is well understood.

Supernova Types

There are two basic techniques with which supernovae have been studied in the past 100 years; namely, analysis of their spectra and analysis of their light curves. The majority of the data has been derived by analyzing changes in the star's spectra as soon as it goes "bang" and in the weeks and months after discovery (see Chapter 8 for more on spectroscopy). In rare cases, data on the progenitor star has been available (i.e., the star has actually been imaged prior to becoming a supernova). The classic case of this was supernova 1987A in the Large Magellanic Cloud. All this data has given astronomers a good idea of how supernovae work, although there are still plenty of puzzling issues. It should, perhaps, be pointed out that there is no danger of our own sun "going supernova." It is not a binary star and it is certainly not a massive star. Our own sun (see Figure 1.2) has an equatorial diameter of 1.39 million km (or 4.6 light-seconds) and is situated at an average distance of

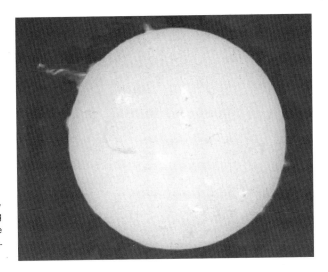

Figure 1.2. Our sun, imaged by Ray Emery using a small amateur telescope fitted with a 40-mm aperture H-alpha filter.

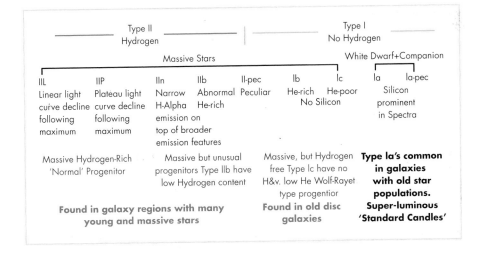

Figure 1.3. Different types of supernovae.

150 million km (500 light-seconds) from the earth. The massive and supermassive stars responsible for most supernovae can have outer regions that would easily engulf the earth's orbit around the sun and are more than a light-hour in diameter! Billions of years from now, our sun will swell up, run out of hydrogen, and cool down. In massive stars, however, this process happens in millions of years and the end result is a supernova, not a slow decline. Massive stars live fast, die young, and go out with a bang! Only the so-called Type Ia supernovae are *not* the result of a massive star collapsing, and they live in binary star systems.

At the simplest classification level, supernovae boil down to two types, labeled Type I and Type II. This spectral classification system roughly boils down to the following: Type I have no hydrogen but Type II do, and the latter look vaguely "sunlike" at first spectral glance. However, Figure 1.3 reveals that things are a bit more complicated.

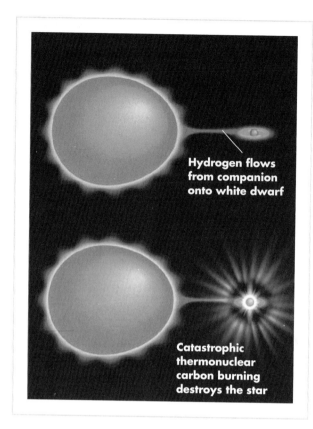

Figure 1.4. Type Ia supernovae, and cataclysmic variable stars, originate in systems containing a white dwarf and a companion star. As described in the text, if the white dwarf is close to the Chandrasekhar limit, the extra hydrogen from the companion can trigger catastrophic thermonuclear carbon burning and the death of the white dwarf.

Type I (hydrogen-free, if you like) supernovae can be subdivided into three further categories; namely, Ia, Ib, and Ic.

Type Ia supernovae are highly prized discoveries because, as we shall see later, they can be used as distance markers throughout the Universe, especially if they are caught on the rise to maximum brightness. Contrary to all other supernovae, Type Ia's are *not* the result of the explosion of a massive star but of a relatively small star. The spectra of Type Ia supernovae show strong evidence of silicon, as well as sulfur and magnesium, and these supernovae are discovered in all galaxies and even in the cores of spiral galaxies where huge stars do not exist. These factors point to Type Ia supernova being the result of the explosion of a white dwarf star in a binary system comprised of the white dwarf and a younger red giant star. Neither of the stars involved in this system will be a massive star; indeed, the white dwarf cannot have a mass of more than 1.4 times that of our sun (1.4 solar masses) at the time of the explosion.

Here is what astronomers think happens when a Type Ia supernova goes bang (see Figure 1.4). Imagine a binary star system in which both components are incredibly close and orbiting each other in an absurdly short period of time; typically a few hours! The more massive star will evolve more quickly and, in the "retirement" phase of its life, will have completely shrunk down to become a white dwarf. Eventually, the less massive star will expand to become a red giant, and some of its hydrogen will get gravitationally sucked down onto the surface of the white dwarf (the popular term for this is *accretion*).

Now, white dwarf stars of, typically, similar mass to our sun have some very unusual properties. Although they may shrink down to Earth size, they have not shrunk far enough to allow carbon-burning nuclear reactions to take place. (Stars use hydrogen as fuel in their normal, healthy state, but as massive stars age they can increasingly use heavier elements still remaining in the star to sustain the nuclear process, rather misleadingly called "burning"). These white dwarves may glow with a brightness of perhaps seven or eight magnitudes less than they once did, be composed of carbon and oxygen, and be incredibly dense, but further contraction is being prevented by the pressure of fast-moving electrons in their centers. This pressure is called *degeneracy* and is created by a quantum effect called the Pauli Exclusion Principle. The degeneracy pressure prevents further contraction regardless of temperature, and the material is described as electron-degenerate. Left on its own, a white dwarf will simply cool down and, after many billions of years, become a cold black dwarf. However, in the binary system we are considering, the white dwarf is not left on its own. Hydrogen flowing from the red companion onto the surface of the electron-degenerate white dwarf burns from hydrogen to helium and adds to the mass of the white dwarf. There is a critical stellar mass here called the Chandrasekhar limit. If a white dwarf exceeds this limit of 1.4 solar masses, the incredible pressure maintained by degenerate electrons will actually be exceeded by the gravitational force and the star will collapse to a neutron star (essentially a monstrous atomic nucleus, maybe 10 km across) or even, if the remaining mass still exceeds 2 to 3 solar masses, in bigger stars, a black hole. However, in Type Ia supernovae, we are only considering what happens if the extra hydrogen from the red giant pushes the white dwarf into a critical state close to the Chandrasekhar limit. At the crucial point, the white dwarf starts to collapse, which heats the carbon in its core and triggers a colossal outpouring of energy that blows the white dwarf apart. The technical term here is *catastrophic thermonuclear carbon burning*. The star never makes it into the neutron star phase, it is blown to smithereens before it can exceed 1.4 solar masses. There are other Type Ia scenarios, too. One involves the coalescing of both binary stars, causing a similar outcome. A third scenario involves the accretion process triggering runaway helium burning just beneath the white dwarf's surface. An asymmetric explosion then triggers the star's destruction. A classical nova eruption involves a similar process (i.e., material from a companion accreting onto the surface of a white dwarf), but in this case, and in dwarf novae/cataclysmic variables (where an accretion disk is involved), the star is not destroyed. It lives on to fight another day and to outburst again. In a supernova explosion, of whatever type, the outburst is colossal and, in Type Ia events, the star is destroyed. In passing, it is worth mentioning that so-called silent supernovae can also occur, where the white dwarf star simply collapses to a neutron star without any fuss. In these cases, there is no carbon and no catastrophic thermonuclear carbon burning. It is the runaway reaction caused by carbon burning that destroys the white dwarf. Remarkably, the nearby secondary star in a Type Ia supernova explosion actually survives! A few hours after the white dwarf detonates, its companion star (probably a red giant like our sun will become in old age) feels the force of the explosion. Within days, the shock wave will have affected every part of the companion star and weeks later half of the companion's mass (i.e., its outer atmosphere) will have been blown into space. However, the companion's core will (probably) have survived.

So just how violent are Type Ia supernovae? Well, typically, they attain an absolute magnitude of −19 or −20. Our sun has an absolute magnitude (i.e., its brightness as seen from 10 parsecs,[1] or 32.6 light-years) of +4.8. As five magnitudes represent a 100-fold brightness increase, we can see that a Type Ia supernova shines briefly as brightly as billions of stars like our sun. Imagine a huge cube in space filled 1,000 high by 1,000 wide by 1,000 deep with stars like our sun. Indeed, there have been cases where such supernovae in small galaxies have outshone the entire galaxy as viewed from the earth. Largely because of their extraordinary brightness, Type Ia supernovae account for roughly half of the supernovae discovered each year. When you consider that a single white dwarf, with no hydrogen accreting onto it, might have a rather feeble absolute magnitude of +12 or so, the brightness of a Type Ia supernova at maximum is even more amazing. Absolute magnitudes can sometimes confuse the beginner, especially where large distances are involved. However, you just have to think in terms of multiples of 32.6 light-years. For example, if a magnitude −19 Type Ia supernova is in a galaxy 32.6 *million* light-years away, then it is a million times farther than the absolute magnitude calibration distance of 32.6 light-years. Brightness falls with distance squared, so it will be a million squared (a trillion) times fainter than −19, or 30 magnitudes fainter. So it will shine with a magnitude of −19 + 30 = magnitude 11; in other words, the brightness of a Type Ia in a very close Messier galaxy.

So, we now know that Type Ia supernovae are, broadly speaking, stars in binary systems that have not quite reached the 1.4 solar mass limit but have suffered annihilation by catastrophic thermonuclear carbon burning. So, presumably, Types Ib and Ic are similar? *Wrong*. Remember, the difference between Types I and II just tells us mainly about the lack of, or presence of hydrogen in the spectra, not whether they are binary or single stars. In fact, Type Ib and Ic supernovae are physically more similar to Type II supernovae, which is why we will now have a look at them first.

Type II supernovae (see Figure 1.5) are thought to be the result of the destruction of a single massive star after the collapse of its core. Types II, Ib, and Ic supernovae are only found in the arms of spiral galaxies. This immediately suggests that they are all associated with massive, relatively short-lived stars, because that is where such stars are found. What do we mean by a massive star that may evolve into a supernova? Well, typically, astronomers class stars of between 8 and 100 solar masses in this category but 20 to 30 solar masses is probably more common.

A really massive star will move through its lifetime very quickly, spending only tens of millions of years or even just a few million years (for the most massive stars) on the so-called main sequence. The main sequence is illustrated in Figure 1.6, and that term has been inextricably linked to the Hertzsprung–Russell (HR) diagram ever since it was first used by Ejnar Hertzsprung and Henry Norris Russell in the second decade of the 20th century. Essentially, the HR diagram is a graph of stellar luminosity (vertical, i.e., the y axis) versus stellar temperature (horizontal, i.e., the x axis). Stars in the top left are hot and bright; stars in the top right are cool and bright. Stars in the bottom left are hot and dim; stars in the bottom right are cool and dim. Our sun is rather cooler than the midrange and a bit on the dim side. The majority of stars live on a line stretching from the top left to the bottom right of the HR diagram. This is the main-sequence path stretching from hot,

---

[1] A parsec is the distance at which the earth–sun distance (149.6 million km) would span the tiny angle of a second of arc (1/3,600 of a degree). That distance is 30.86 trillion km, or 3.26 light-years. It is an abbreviation of "parallax of one arc-second," related to the earth–sun distance.

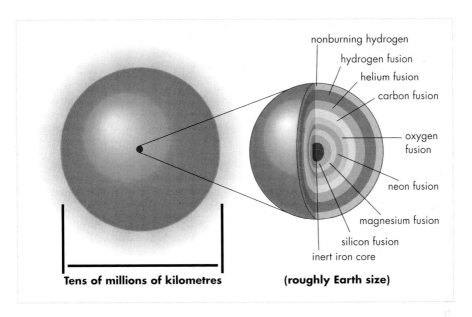

nonburning hydrogen
hydrogen fusion
helium fusion
carbon fusion
oxygen fusion
neon fusion
magnesium fusion
silicon fusion
inert iron core

Tens of millions of kilometres   **(roughly Earth size)**

Figure 1.5. As described in the text, in all other supernovae except Type Ia's, the progenitor star is a massive or supermassive star that burns through increasingly heavy elements before it runs out of fuel. Although the massive star may appear to have a diameter of tens (or hundreds!) of millions of kilometers, the main nuclear reactions involving the heavy elements are only taking place within a central Jupiter-sized region, and the iron core is typically about the same size as the earth.

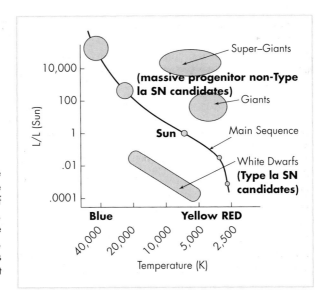

Super–Giants

(massive progenitor non-Type Ia SN candidates)

Giants

Main Sequence

White Dwarfs
(Type Ia SN candidates)

Sun

L/L (Sun)

10,000
100
1
.01
.0001

Blue   Yellow RED

40,000   20,000   10,000   5,000   2,500

Temperature (K)

Figure 1.6. Most stars live on a line stretching from the top left to the bottom right of the HR diagram shown here. This is the main-sequence path stretching from hot, bright, massive bluish stars to cool, dim, lightweight reddish stars.

bright, massive bluish stars to cool, dim, lightweight reddish stars. However, not all stars live on this line. Giants and super giants (non Type Ia supernova candidates) live on the right and top right and white dwarves (Type Ia candidates if in a suitable binary system) live in the lower region.

Unlike the white dwarf end-state of stars of less than eight or so solar masses, heavier stars have a different fate. At this point, the reader may spot a discrepancy. Surely the white dwarf end-state can only exist for stars lighter than the Chandrasekhar limit of 1.4 solar masses? Actually, no, because in the later life of giant stars that swell up to become a supergiant star, 80% of the unburned hydrogen can be expelled, reducing the mass to just under the Chandrasekhar limit. However, in truly massive stars, as the star runs out of hydrogen to burn and then runs out of helium fuel, too, the temperatures become high enough (600 million K) for the next layer of material, carbon, to burn in a nuclear reaction. This carbon-burning phase may last 1,000 years, and in the sort of stars we are considering (greater than 8 solar masses) the core temperature may escalate to 1 billion K. The next layer down (i.e., neon), will burn at that temperature, but, typically, only for about a year. At around 1.5 billion degrees, oxygen can burn, too. The burning, or nuclear fusing, of carbon, neon, and oxygen lead to the production of other elements detected in the spectra of supernovae (i.e., sodium, silicon, and magnesium). At the phenomenal temperature of 3 billion degrees K, silicon can sustain the stars burning process, too, albeit for a pitiful length of time, namely a few days at best. (The term *burning* may conjure up images of a cosy log fire here. In reality, we are talking about thermonuclear reactions of course.) As long as thermonuclear processes are taking place, the energy produced helps prevent the star collapsing gravitationally under its own weight. It is a case of gravity versus nuclear reaction pressure. The silicon burning process results in the formation of iron, and the core of a giant star close to the end of its life will also contain nickel and cobalt, converted from the magnesium, silicon, and even sulfur burnt in the ageing star's last days, but the game stops there. You need to put in more energy than you get out before iron will fuse, so after a day or two of silicon burning and (probably) a matter of hours after burning stops and the dormant iron core is finalized, the star collapses and a Type II supernova is born. However, core collapse supernovae can produce smaller amounts of really heavy elements, like gold, when they go bang. Just prior to collapse, theory predicts that the iron core will have a density of 500 million tons per cubic meter and a temperature of maybe 8 billion degrees K!

The fundamental point here, as we saw with Type Ia white dwarves, is that if the iron core of a fuel-exhausted massive star is greater than 1.4 solar masses, electron degeneracy will not stop it collapsing. As I mentioned earlier, it will collapse to become a neutron star or a black hole. But what actually happens at the instant of core collapse? Even in the 21st century, the precise details of the collapse of a massive star are not fully understood. The collapse takes place before the supernova flares visually and, regardless of that, spectroscopes cannot penetrate beneath the star's surface. So much depends on the precise mass of the star, too. However, events are thought to proceed roughly as follows. Within a fraction of a second of the core collapse starting (yes, you did read that right . . . a fraction of a second!), the iron core's electrons and protons are crushed together to form neutrons, which have no electrical charge. Stupendous numbers of neutrinos are released in the process and the core shrinks to around 10 km in radius in less than a second. The matter is so dense in the resulting neutron star that the proverbial teaspoonfull of it would weigh hundreds of millions of tons. This really hammers home to me how empty normal matter is, with huge chasms between atomic nuclei. Astronomers think that the neutron star will avoid collapsing further (i.e., becoming a black hole), if the remaining mass is less than about 2.5 solar masses. So-called neutron degeneracy pressure will prevent the collapse. Of course, the giant star did not just consist of the iron core; the whole thing is shrinking in the collapsing process. As the lighter non-iron elements surrounding the core hit the giant 10 km diameter atomic nucleus, they are thought to rebound. This rebounding shock wave blasts the rest of the star away. The current thinking is that the massive neutrino outpouring plays a big part in this process, too, but the precise role of colossal convection currents and magnetic fields in the process are still a subject for considerable debate. Neutrinos usually pass through ordinary matter as if it was not there, but astronomers think that in the bizarre environment of a collapsing stellar core, they may play a part in the supernova explosion, conveying perhaps 10^{46} joules of energy at light speed away from the core. (No, that is not a misprint. It is 1 followed by 46 zeroes!) Depending on the exact size of the giant star, it may take hours for the shock wave to reach its visible surface, although a figure of 30 minutes is a favorite one amongst astrophysicists. This time period indicates the gap between the detectability of a neutrino burst and the star brightening. Of course, only supernova 1987A in the Large Magellanic Cloud had an associated, detected neutrino burst. Remember, although the core may now be only 10 km across, and with a density of 3×10^{14} tons per cubic meter, the giant star may well have been tens of millions of kilometers in diameter prior to its collapse. Essentially, the outer layers of the star are blissfully unaware of the maelstrom that is propagating out from the center at initial speeds as high (perhaps) as 1/10 of the speed of light. When the shock wave reaches the star's outer layers, the brightness of the star increases dramatically: the supernova is born. At this point, all of the material above the core, maybe 20 or 30 solar masses or more, will be flung out into space at speeds of up to 6% or 7% of the speed of light. Although not as bright as Type Ia supernovae, Type II (and most non–Type Ia) events typically peak at an absolute magnitude of −17 (i.e., equivalent to hundreds of millions of suns). A typical, massive, Type II progenitor (if the word *typical* can be applied) of 10 to 20 solar masses might have a pre-explosion absolute magnitude of around −7. In other words, these were 50,000 sun power monsters that brightened by "only" 10,000-fold on going supernova. Because of the predetonation brightness of these giant stars, there is always a slim chance that a deep galaxy image prior to the explosion (e.g., a Hubble image) might show the original star. No such chance exists with the tiny, but ultimately awesome, Type Ia progenitors.

Type Ib and Ic supernovae are also believed to involve the core collapse of a giant star, but, as we have seen, being Type I they, by definition, contain no hydrogen. The best theory to explain this is that the stellar winds from these very massive stars have blown the outer layers/hydrogen away prior to the supernova explosion. This effect can happen in so-called Wolf–Rayet stars or in binary star interactions. The core collapse supernovae do, at least, leave a remnant of their existence (i.e., a neutron star or a black hole). However, it is thought that Type Ia events utterly destroy the white dwarf star. As we have seen, the spectra of Type Ia supernovae show evidence of silicon, sulfur, and magnesium. The core collapse, but hydrogen-free, supernovae of Types Ib and Ic do not exhibit prominent silicon lines but do show oxygen and magnesium lines and, in the case of Type Ib, they show helium lines, too. Thus, Type Ic supernovae have no hydrogen and virtually no helium. The spectra of Type II supernovae, by definition, show hydrogen lines and similar spectra to our own sun, (i.e., the so-called solar abundance). Type II supernovae can be further subdivided into categories IIb, IIn, IIL, and IIP as follows. Type IIb are the abnormal, helium-rich supernovae with much hydrogen removed by tidal winds, whereas the more normal case is the massive, hydrogen-rich progenitor.

Type IIn supernovae contain hydrogen and have narrow H-alpha emission lines seen on top of broader emission features in the spectra. The later light curve of Type II-n supernovae tends to be slow. The more normal hydrogen-dominated Type II supernovae break down into Types II-L and II-P.

The difference between these L and P subcategories can almost be guessed at from the letters. L denotes a linearly fading light curve after maximum, whereas P indicates that a temporary plateau is seen. Obviously, this precise classification cannot be deduced at maximum brightness. An extra term sometimes added to designations is p or pec, standing for peculiar. Thus, SN 1991T in NGC 4527 (Figure 1.7) is sometimes designated as Type Ia-p or Ia-pec. A Type II-pec category was proposed by the astronomers Doggett and Branch in 1985 to contain all of the old-style Fritz Zwicky categories that he had labeled as Types III, IV, and V.

Figure 1.7. The very bright and peculiar Type Ia supernova 1991T in NGC 4527. Image by the author.

Figure 1.8. The Type Ia supernova 1994D, in the galaxy NGC 4526, shines brightly in the lower left of this Hubble Space Telescope image. Image: Space Telescope Science Institute/NASA/High Z Supernova Search Team.

These categories are now defunct, and essentially the modern Type II-pec contains all the Type IIs that do not fit into the aforementioned four categories of L, P, b, or n.

Type Ia supernovae are most common in galaxies with old star populations. As we have seen, a white dwarf in a binary star system is thought to be the progenitor and not a collapsing giant star. Type Ib supernovae are rare animals and seem to occur mainly in old disk galaxies. Large, young stars that have lost their hydrogen are thought to be responsible. The so-called Wolf–Rayet type stars are, again, a likely contender. Type Ic supernovae are very rare objects indeed containing no hydrogen and no helium either! Their progenitor stars are probably similar to those of Type Ib. Type II supernovae (i.e., all those with hydrogen in the spectra), are common in galaxies/galaxy regions containing a lot of young massive stars. Figure 1.8 shows the Type Ia supernova (SN) 1994D in NGC 4526, imaged by the Hubble Space Telescope.

Chapter 2

Supernovae to Measure the Universe

Astronomers can measure the distances to far-away objects if they know the precise brightness of the object they are studying. If you know how bright a tiny star-like point looks and also how bright the sun responsible really is, in absolute terms, you can calculate the distance. Obviously, the farther away one of these *standard candles* is, the fainter it will appear. Move it 10 times farther away and it will look 100 times fainter. Type Ia supernovae are the ultimate standard brightness candles. Take a look at the remarkable 2005 Hubble image of the galaxy NGC 1309 in Eridanus in Figure 2.1. This galaxy is 100 million light-years away and its distance is close enough to be measured using standard Cepheid variable stars (whose period of variation is linked to their brightness). NGC 1309 is also close enough to produce a supernova within an amateur astronomer's detection range. In 2002, the 15th magnitude supernova 2002fk erupted in this galaxy. Fortunately, it was a Type Ia (i.e., a superluminous standard candle appearing in a galaxy whose distance had already been determined). Thus, astronomers are doubly confident that it lies at 100 million light-years. When amateurs discover supernovae in such relatively nearby galaxies, it provides a useful check on the Type Ia distance/luminosity scale. But look at the huge number of fainter background galaxies in Eridanus, surrounding NGC 1309. Many of these are dozens of times farther away and thus only Type Ia supernovae (not Cepheids) within them could possibly pin down their distance.

It was in the early 1990s that information from the Calán/Tololo Supernova Survey (in Chile) was studied in detail and revealed that Type Ia supernovae, the exploding white dwarves in binary systems, were predictable enough in brightness to be used as crude standard candles effective across hundreds of millions of light-years. The crucial factor that made these explosions usable was the fact that astronomers deduced that bright Type Ia supernovae lasted longer than faint ones. It was possible therefore to use the duration of a Type Ia outburst to correct for the differences in their brightness at maximum. Being so extraordinarily bright in real terms (i.e., as bright as a small galaxy) gave astronomers an incredibly powerful ruler to estimate the distance to far-away galaxies. It might be thought that all Type Ia explosions would be the same brightness; after all, when one of these white dwarves, containing carbon, is pushed close to 1.4 solar masses due to infalling hydrogen from its companion, it goes bang. How can there be much variation? In fact, every star is different, the exact

Figure 2.1. NGC 1309 in Eridanus. This is a galaxy 100 million light-years away, imaged by Hubble in 2006. Note the myriad of more distant background galaxies.
Image: NASA/ESA/Hubble Heritage/Aura/STSCI.

carbon content is different, and every binary system is different, too. The only fixed factor is that the white dwarf cannot exceed 1.4 solar masses. However, in practice, the initial, uncorrected absolute magnitudes of almost all Type Ia supernovae fit between −19.0 and −20.0 (i.e., a factor of 2.5 times in absolute brightness). The majority only have a spread of 30% to 40% and once the "brighter means longer-lived" correction factor is applied to the light curve, a really useful standard candle emerges with roughly 10–20% accuracy. This does not mean that there are no exceptions though. Roughly a third of Type Ia supernovae are unusual in one way or other, and there are a small number that are wildly under- or overluminous. The overluminous events are very hard to explain without invoking vast quantities of nickel-56 in the core.

Needless to say, in this era of giant Earth-based telescopes (like the Kecks, the VLT, and Gemini), not forgetting the smaller but optimally placed Hubble Space Telescope, astronomers are keen to push the Type Ia standard candle to the limit. Just how far across the visible Universe can such brilliant events be seen? Well, Figure 2.2 shows the farthest supernova detected. SN 1997ff in an anonymous galaxy in the so-called Hubble Deep Field is estimated to have a z of 1.7 corresponding with an age of (probably) some 11.3 billion years. Figure 2.3 shows an image of the even deeper Hubble Ultra-Deep Field.

At this stage, I think we need to familiarize (or remind) ourselves of the confusing concepts of distance, redshift, and look-back time, which are involved in understanding an expanding Universe. Ever since the Big Bang occurred, roughly 14 billion years ago (at least that is the most popular figure), everything in the Universe has been expanding away from everything else. Unlike the laymen's concept

Figure 2.2. Supernova 1997ff in an anonymous galaxy in the so-called Hubble Deep Field is estimated to have a record-breaking z of 1.7 corresponding with an age of (probably) some 11.3 billion years. Image: NASA/STSCI.

Distant Supernova in the Hubble Deep Field
Hubble Space Telescope • WFPC2

Difference: 1997-1995

NASA and A. Riess (STScI) • STScI-PRC01-09

Figure 2.3. The deepest astronomical image ever taken: the Hubble Ultra-Deep Field. Exposures were made between September 24th, 2003, and January 16, 2004, during 400 individual orbits of the Hubble Space Telescope. The total cumulative image time was a million seconds, or 12 days. The picture is of a region in the constellation of Fornax. Some 10,000 galaxies are visible in the image, some as young as 400 million years after the Big Bang, and fainter than magnitude 30. The field is approximately 3 × 3 arc-minutes. Image: NASA/ESA/STSCI.

Hubble Ultra Deep Field
Hubble Space Telescope • Advanced Camera for Surveys

NASA, ESA, S. Beckwith (STScI) and the HUDF Team

STScI-PRC04-07a

of this as being an explosion from a single point source that marked "the middle," cosmologists prefer the stretching rubber sheet analogy. In other words, every point on the sheet is expanding away from every other point, and wherever you live, you see the same thing: objects moving away from you. If you look across our solar system, or even our galaxy, the effects of this recession are negligible. But as you look at very distant galaxies, you can see that they are moving away from us at speeds that are a significant fraction of the speed of light! This is deduced by the fact that the lines in their spectra are redshifted (i.e., the wavelength increases and moves from the visible region into the infrared, and beyond). These galaxies are, or were, typically, hundreds of millions of light-years away when the light left them, and the time the light has taken to reach us is known as the look-back time. It is important to recognize that one has to be very careful when stating that an object is at a specific distance. A galaxy will have been at a certain distance from us when the light we see left it. The light will have taken a certain time to arrive at the earth and, by now, the galaxy will be much further away from us. So what distance or time units should we use to describe how far away the object was or is? Astronomers most frequently use the term z to represent the redshift of an object at huge astronomical distances. In fact, they use the term far more often than any distance estimate, simply because it is derived precisely from the observations of an object's spectrum and is not subject to any uncertainty. The term z is a simple concept, namely:

$$z = (\text{Observed wavelength} - \text{True wavelength}) / \text{True wavelength}$$

In terms of simple, low-speed Doppler shift, this is equivalent to the galaxy's velocity away from us (v) divided by the speed of light (c). However, things are not quite that straightforward with objects moving at velocities that are a significant fraction of the speed of light. Values of z in excess of 1.0 would indicate recessional velocities faster than the speed of light if not for the effects of relativity. In fact, objects with a much higher z value than 1.0 are regularly imaged by professional astronomers (i.e., lines in the spectra of these objects have been shifted by a factor of 2 or more in wavelength. When relativity comes into play, the formula $z = v/c$ becomes $z = \sqrt{\dfrac{(c+v)}{(c-v)}} - 1$.

So when an astronomer describes an object as having a certain z value, how can we interpret that into a meaningful distance value? The truth is that the distance aspect, like so many things in cosmology, is an educated piece of guesswork. Indeed, the very reason distant supernovae are so useful is that they can be used to refine distance calculations and the rate of expansion of the universe. Astronomers prefer to use look-back time as the most meaningful phrase in this context and, of course, as the speed of light is known, and, by definition, is one light-year per year, a distance can be inferred from this. However, there are two variable factors that prevent a precise relationship between redshift and any sort of distance estimate. The first factor is called the Hubble Constant and it tells us how fast the Universe is expanding as one looks further into space. The units are kilometers/second (speed) per megaparsec (distance), where a megaparsec is 3.26 million light-years. Astronomers currently believe this figure (which has been constantly refined over the past 100 years) is close to 70. In other words, if the redshift of a galaxy indicated it was travelling at 70 kilometers per second, the corresponding look-back time would be 3.26 million years and the distance would be 3.26 million light-years. However, 3.26 million light-years is next door to our own galaxy and, in practice, galaxies hundreds and thousands of times farther away than this are being studied. Plus, remember, in the time a galaxy's light has taken to get to us, especially a distant galaxy receding from us at a large fraction of the speed of light, that galaxy will have moved much farther away. All things considered, the faster an object appears to be receding from us, the more meaningless is our concept of "distance," hence astronomers' preference for the z, or redshift, term. This preference is reinforced when we consider the second variable factor.

If one assumes the Universe has always expanded at the same rate (i.e., 70 kilometers per second per megaparsec), it is only necessary to pin down the exact value of the Hubble Constant H_0 to determine how far away objects are (or were). However, the Universe has a mass, and mass exerts a gravitational force. One would expect this to slow down the rate of expansion of the Universe. If you fire a missile vertically up into the air, at high speed, it will eventually succumb to the force of gravity and slow down and then crash back to Earth. If you fire it at more than 40,000 kilometers per hour, it will escape from Earth completely. If you fire it at precisely 40,000 kilometers per hour, it will get slower and slower until, many days later, its vertical speed would stop dead as its velocity and the diminishing force of gravity cancelled out. (Actually, in practice, this would not happen, because, in the first instance, the moon would have an effect, as would the sun and the other planets . . . but I hope you get my reasoning!) Astronomers face a similar situation with respect to the Universe. Our second variable factor is simply the change of the Hubble Constant with time. Surely one would expect the expansion of the Universe to slow down as gravity hauls everything back together? The terminology used here is whether the Universe is open (and expands for ever, but the expansion slows down), flat (i.e., after countless billions of years everything just stops dead), or closed (ultimately, gravity hauls everything back in a "Big Crunch"). The term *Omega* (or Omega$_M$) is used to represent the ratio of the actual mean density of the Universe to the critical "flat" density where Omega = 1. An open Universe will have Omega <1 and a closed, Big Crunch Universe will have Omega >1. Obviously, different Omega values will affect the past and future values of the Hubble Constant. From all of the observations to hand, and including the effects of the mysterious dark matter that cannot be seen (!), astronomers think that Omega is close to 1, or at least between 0.1 and 2.0. It may even be 1.0, which would be astonishing, like the aforementioned missile stopping, as velocity and gravity precisely cancel!! When one considers that Omega could be any random value, why is it so close to 1.0?

But before you relax at having digested the Hubble Constant and the relative critical density of the Universe, Omega, there is a third factor in the equation that adds another twist to the tale. This third factor is called the Cosmological Constant and was a term originally added to the field equations of general relativity by Albert Einstein in 1917. At that time, the Universe was thought to be static (i.e., not expanding). So, Einstein invented a term that would counteract the gravitational force and prevent everything from getting pulled together; a sort of "anti-gravity" term if you like. When Hubble discovered that galaxies were receding, Einstein described his own Cosmological Constant as the greatest blunder of his life, because it was conceivable that he might have deduced that the Universe was expanding if he had not invented the term. However, the Cosmological Constant is back in fashion. Why? Well, largely because of the analysis of distant Type Ia supernovae. With so much effort in astrophysics being directed at finding out how old the Universe is and how much matter it contains, especially with much of this matter being dark (i.e., invisible, unlike stars), Type Ia supernovae are an invaluable tool. By studying their spectra and light curves, their absolute brightness and recessional velocity can be measured. This immediately allows astronomers to deduce what the recession speed of a distant supernova was in much earlier times. After all, the distance the light has travelled can be deduced from how bright it appears, and the speed it is (or, rather, was) receding from us can instantly be derived from its spectra. In the late 1990s, measuring the redshift of distant Type Ia supernovae became a major project and two international teams dominated the field. They were called The Supernova Cosmology Project and The High-z Supernova Search Team. Obviously, a major challenge was simply discovering enough distant supernovae, some around 8 billion or 9 billion years old, to measure. A further challenge was obtaining spectra of such faint objects whose normal spectral lines had been shifted far into the infrared. A nearby 15th mag Type Ia supernova at a distance of 100 million light-years or so will only have a z of approximately 0.01. Amateurs with 25- to 35-cm apertures rarely discover supernovae much fainter than about mag 18 (i.e., objects with a z greater than 0.07 or so, or a distance limit approaching 1 billion light-years). By the time we get to Type Ia supernovae with a seriously high z of 1.0, the target has faded to a paltry 25th magnitude. However, despite their faintness, even before the year 2000 the dozens of discovered high z supernovae that were studied showed an extraordinary trend. The oldest supernovae (with z's as high as 1.2) were definitely fainter than expected from their redshifts. This meant that they were farther away and that these very distant supernovae, and their galaxies, were receding away from us less quickly than expected. In other words, many billions of years ago, the universe was expanding more slowly. But in more recent aeons the expansion of the Universe has been accelerating, not slowing down due to gravitation! Figure 2.4, by Saul Perlmutter, shows how the old results from the Calán/Tololo Supernova Survey merge with the more recent and unexpected results on distant supernovae made by the Supernova Cosmology Project and the High-z Supernova Search Team.

Currently, the only explanation for this "recent" acceleration is a repulsive force (opposing the attractive force of gravity), that is, something similar to Einstein's Cosmological Constant. Astronomers are now referring to this repulsive force as dark energy. Maybe Albert was right in the first place. Remember, dark matter is the material we cannot see in the Universe, but we know it is there by studying the rotations of galaxies and superclusters and detecting its gravitational pull. If dark matter were not there, these galaxies and superclusters would fly apart. Dark energy is working on a much larger, universal scale and in the opposite direction to gravity. Recent work has studied even higher z supernovae (up to $z = 1.7$) that occurred 10 billion years ago and has reinforced the view that the Universe was then decelerating, as expected, after the Big Bang (which occurred roughly 14 billion years ago). Current thinking, in 2006, is that the Universe started seriously accelerating about 5 billion years ago ($z = 0.5$), as the mass in the Universe had then thinned out enough for the steady Cosmological Constant/dark energy force to dominate. Although cosmologists are still perplexed as to what dark energy (and dark matter) are, the greatest tool for verifying all this is still the Type Ia supernova at great distances. Figure 2.5, by Saul Perlmutter, explains all this in a very useful graphical form, although even this graph takes some time to get your head around.

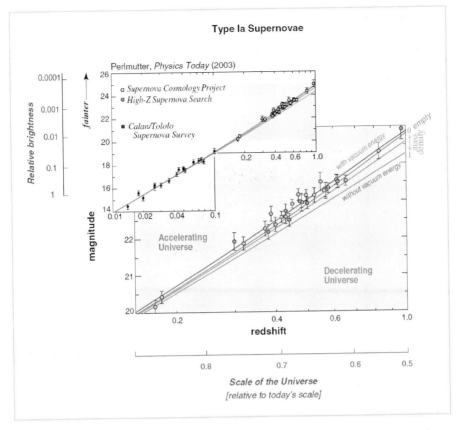

Figure 2.4. Observed magnitude versus redshift for well-measured distant and (in the inset) nearby Type Ia supernovae. The older Calan/Tololo Supernova Survey results merge with the more recent and unexpected results on distant supernovae made by the Supernova Cosmology Project and the High-z Supernova Search Team. Diagram: By kind permission of Saul Perlmutter (originally reproduced in *Physics Today*, 2003). (see color plate)

Is the expansion of the Universe really accelerating? A few astronomers still have their doubts. Maybe the high-z supernovae are not abnormally faint just because they are further away; maybe Type Ia supernovae were subtly different in earlier times? Or, and to some this amounts almost to religious heresy, maybe the theory of relativity is wrong and simply does not hold up across the vast times and distances being considered. Even a variable force of gravity has been postulated. Only time will tell.

At the current time there seems to be a consensus amongst most cosmologists that the mass/energy composition of the Universe consists of 75% dark energy, 21% dark matter, and 4% normal matter, in other words, dark energy wins out and the Universe is accelerating. This model not only ties in with the observations of distant supernovae but also with inflationary Big Bang theory, and it has all come together because of the study of those critical Type Ia supernovae.

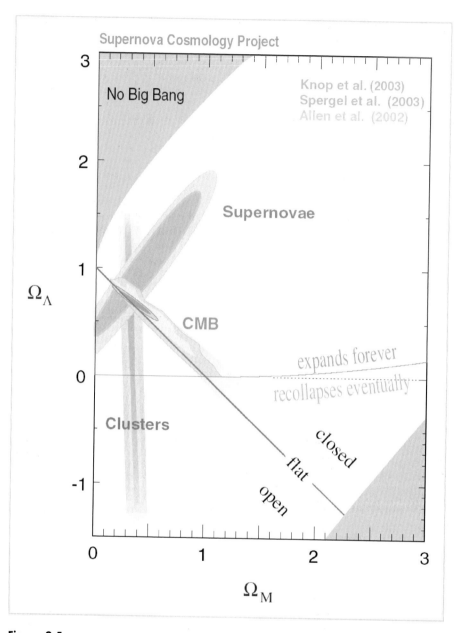

Figure 2.5.

(see color plate)

Figure 2.5. Confidence regions for Omega Mass versus Omega Lambda. This diagram shows the confidence in the cosmological models for the kind of Universe we live in. The models of the 1980s and 1990s have been totally shattered by the distant supernova findings. If we had all the information (e.g., from an all-powerful and knowledgeable being), we could stick a pin on a specific point on the graph and say "this is how our universe is." The x axis represents the relative mass density of the Universe, Omega M (i.e., how much gravity producing potential there is). Obviously, the dark matter content is of crucial importance here. The y axis represents the acceleration parameter, or rather the Cosmological Constant's relative density, or Omega Lambda, which is now associated with the term dark energy and the accelerated expansion of the Universe. Note, this is Omega Lambda, not the Cosmological Constant itself, but the ratio of the "density" of the cosmological constant to the critical zero point. In a theoretical Universe where, eventually, the Big Bang expansion will be exactly halted by the gravitational influence of all the matter, the Omega Lambda value on this graph would be zero. The critical supernova ellipses in the mid/top left of the diagram show the Universe model confidence contours based on ultradistant Type Ia supernova redshift and brightness measurements. These contours illustrate the amazing result that the Universe is still accelerating (and faster than in the past). At the time of writing, a y-axis acceleration parameter of around 0.7 and an x-axis relative mass density of around 0.3 seem most likely, especially when combined with Cosmic Microwave Background measurements (CMB on the graph) and measurements of the dark matter content in galaxy clusters (Clusters on the graph). Essentially, this means that the Universe started with a Big Bang, will expand forever, and, also, is essentially flat. By flat we mean that the fabric of space obeys Euclidean geometry and that a small chunk of the whole cosmos expanded very rapidly in the first instant after the Big Bang to form the almost perfectly smooth background radiation of the observable Universe we see today. Mathematically, a flat Universe is one in which the relative density parameters Omega M plus Omega Lambda equal one. If there were no need for a Cosmological Constant (as cosmologists imagined less than 10 years ago), this flat point would occur at the point where Omega M was unity and Omega Lambda was zero. The strongest evidence for flatness comes from WMAP (the Wilkinson Microwave Anisotropy Probe) CMB measurements, and the evidence that our Universe is flat constrains our Universe to the diagonal line in Figure 2.5. Thus, on current evidence our Universe fits within the lower part of the supernova confidence ellipse where it is crossed by the diagonal line marked flat and where the CMB and Clusters lines also intersect. Prior to the relatively recent supernova evidence that the acceleration of the Universe was increasing, this would all have been considered extremely unlikely, if not preposterous! Because the Universe is thought to be flat, and so Omega M plus Omega Lambda, by definition, equals one, or 100%, the Universe is now often described in the media as consisting of 70% dark energy and 30% (dark matter + visible matter). Diagram: By kind permission of Saul Perlmutter, et al., The Supernova Cosmology Project, based on Knop et al. (2003).

It might be thought, from the above, that only the study of the most distant supernovae could have any importance. However, having a complete understanding of supernovae, even ones much closer, and comprehending their recession speeds at all epochs makes the theories more precise and places more confidence in the whole science of measuring the size of the Universe. Amateur astronomers still have a vital role to play.

Crab Nebula ▪ M1　　　　　　　　　　　　　　　　　　　　　　*HST* ▪ WFPC2

NASA, ESA, and J. Hester (Arizona State University)　　　　　　STScI-PRC05-37

Figure 3.1. The Crab Nebula imaged by the Hubble Space Telescope in 1999 and 2000. This amazing image is a mosaic of 24 individual images. Image: NASA, ESA, J. Hester and A. Loll (Arizona State University). (see color plate)

ated with a naked-eye supernova that was seen in AD 386 (see Figure 3.5). There is some evidence for other naked-eye supernovae in the years 185 and 393, but the 185 event may just have been a staggeringly bright comet, and the AD 393 event does not seem to have been spectacular.

So, remarkably, there have been no visible supernovae in our own galaxy for a staggering 400 years, despite the fact that, statistically, one might expect two or three per century to occur. Of course, statistics are one thing and reality is another. Although it does not affect the probability calculation, it is a sobering thought that even when a supernova does occur in our galaxy, it can take thousands or tens of thousands of years for the light to reach us. If a naked-eye supernova appeared in our skies tomorrow, it may have actually occurred a few hundred years ago or tens of thousands of years ago. But, on average, in the last barren 400 years one would have expected the light from eight, ten, or twelve supernovae to reach us. So what

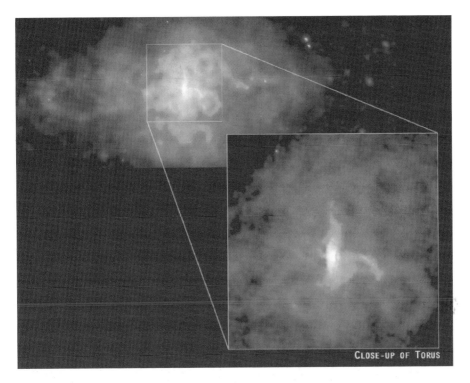

CLOSE-UP OF TORUS

Figure 3.2. The Chandra X-Ray Observatory image of 3C58; the remains of a supernova observed on Earth in AD 1181. In the center, an X-ray jet erupts in both directions and extends over a distance of several light years. Image: NASA/CXC/SAO/S, Murray et al.

Figure 3.3. The picture shows a Chandra X-Ray Observatory image of Tycho's supernova of 1572. This shows an expanding bubble of multimillion-degree debris inside a more rapidly moving shell of extremely high-energy electrons. Image: NASA/CXC/Rutgers/J. Warren and J. Hughes et al.

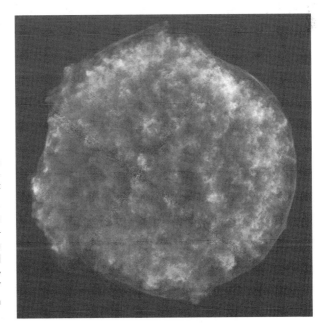

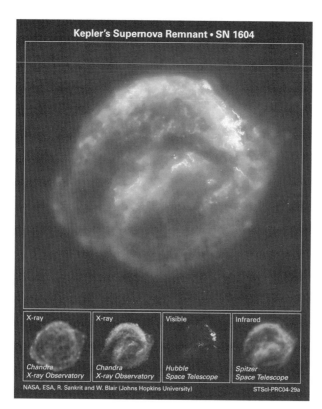

Figure 3.4. The Hubble Space Telescope, the Spitzer Space Telescope, and the Chandra X-Ray Observatory joined forces to take this image of Kepler's supernova of 1604. Image: NASA/ESA/JHU/R. Sankrit and W. Blair.

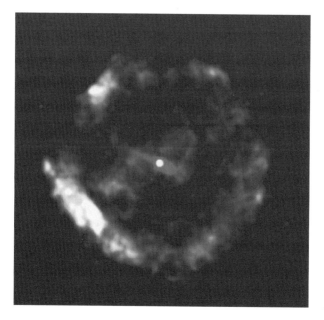

Figure 3.5. This Chandra X-Ray Observatory image shows a pulsar exactly at the center of the supernova remnant known as G11.2-0.3. The evidence from Chandra implies that the pulsar was formed in the supernova of AD 386, which was witnessed by Chinese astronomers. Image: NASA/McGill/V. Kaspi et al.

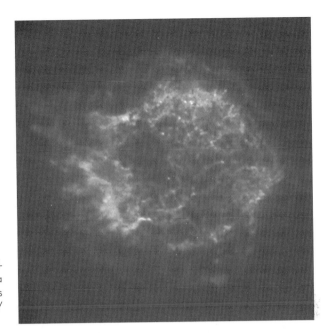

Figure 3.6. A Chandra X-ray image of the supernova remnant Cassiopeia A (Cas A). Image: NASA/CXC/SAO/Rutgers/J. Hughes.

went wrong? If we look at the historical supernovae that we are sure about (i.e., the five mentioned above), we can see that those five appeared in a time span of 600 years. But even that is less than one per century. If one starts at 1006 and counts through to the present day, one millennia exactly (as I type these words), we have one visible supernova every 200 years. This is roughly a quarter of what one would expect and the 400-year dearth since 1604 is most disappointing. So can we explain this shortfall? Well, for starters there was almost certainly a supernova around 1667 or 1680, because it left the X-ray emitter Cas A; the brightest X-ray source in the sky apart from our sun (see Figure 3.6). Interstellar absorption in the plane of the Milky Way must have dimmed the object enough to make it inconspicuous to the casual observer of the time. Professional astronomers estimate that we may only see as little as 10% of the optical regions of our galaxy because of intervening dust. If one inverts this figure of 10%, it might conceivably be argued that we miss ten times as many supernovae as we see. Suddenly things start looking closer to the predicted values. However, we should not dismiss the 400 years since 1604 quite so lightly and, unless theory and reality are completely at odds, we have to put these barren years down to a statistical quirk or a gap in our understanding. Also, let us not forget the supernova in the Large Magellanic Cloud: SN 1987A. Okay, this was not, strictly speaking, in our own galaxy, but it was in a small satellite galaxy nearby. If supernovae in our own galaxy occur every 50 years or so, then we will soon know if the theories fall short. With the array of X-ray and gamma ray equipment now operated by astronomers, it is unlikely that a modern (in the sense that its radiation had just arrived) supernova in the galaxy would escape detection, even if, visually, its light was highly attenuated by galactic dust. So far, SN 1987A is the only nearby, naked-eye supernova to go off in the modern detector era.

There is another aspect to all this, too. Massive stars often leave a pulsar remnant when they go supernova, and many have been detected since the historic discovery of the first four pulsars by Jocelyn Bell Burnell, at Cambridge, in 1967.

Astronomers are pretty sure that pulsars "switch off" after about 10 million years when their magnetic fields have weakened significantly. The number of pulsars detected in our Milky Way neighborhood indicates that there must be roughly 1 million active pulsars in the whole Milky Way Galaxy. If you think about it for a moment, this implies that pulsars from massive supernovae must be born every 10 years or so in the Milky Way, a rate that is puzzlingly far greater than the expected (let alone observed) rate of occurrence of such supernovae in our galaxy. There is obviously still much that we do not yet know.

Supernova 1987A

On February 23, 1987, the brightest supernova for 383 years appeared in the Large Magellanic Cloud (see Figure 3.7), an irregular dwarf companion galaxy to our own Milky Way. The Magellanic Clouds, Large and Small are, sadly, not visible from the United Kingdom, so Northern Hemisphere observers were denied this special treat. At 170,000 light-years away, this supernova was never going to rival Venus, but it was the only naked-eye supernova in the modern detector era, so it was a real treat for professional astronomers. Before-and-after pictures of the region, taken by

Figure 3.7. The very bright star in the bottom right of this image is the supernova 1987A in the Large Magellanic Cloud (LMC): the first supernova to be visible to the naked eye for 383 years! The complex red feature in the top left is the LMC's famous Tarantula nebula. The angular width of this image is about 28 arc-minutes. Image: © 1987 Anglo-Australian Observatory. Photograph by David Malin. (see color plate)

Figure 3.8. The picture shows the field of supernova 1987A in the Large Magellanic Cloud 10 days after the explosion and (on the right) before the explosion. The original progenitor star, Sanduleak −69° 202, can be seen on the right-hand image. Image: © 1987 Anglo-Australian Observatory. Photograph by David Malin. (see color plate)

David Malin, illustrate nicely the violence of the event (see Figures 3.8 and 3.9). Uniquely, neutrinos were detected coming from this supernova, and a progenitor star was identified from images taken before February 23. It came as a bit of a surprise when the progenitor was found to be a blue supergiant star, because prior to 1987A astronomers had assumed that core-collapse progenitors were red supergiants. It is now thought that the star, designated Sanduleak −69° 202 (its declination being 69° South), was a red supergiant up to a few thousand years before the explosion, but then shrank and heated up, thus changing to a "B3I" blue supergiant. Neutrinos from SN 1987A were detected by three neutrino detectors on Earth; namely, the Kamiokande II detector in Japan; the IMB detector in Fairport, Ohio, USA; and the Baksan detector under Mount Andyrchi in the northern Caucusus Mountains, Russia. These detectors are buried deep under rock to filter out the constant cosmic ray bombardments that would otherwise flood the detectors. Essentially, they are huge liquid-filled bodies (3,000 tons of water, a kilometer underground, in the case of Kamiokande II) that provide enough molecules for neutrinos to occasionally react with; at least, when billions of neutrinos are flowing through them. Neutrinos rarely interact with any matter (i.e., the probability of an interaction with an atomic nucleus is almost infinitesimally low). However, when the number of neutrinos is almost infinitely high, as in a supernova explosion, the near-infinite and near-infinitesimal cancel out and a few particle events may be registered by the detectors monitoring the huge masses of underground liquid. In

Figure 3.9. The brightest star in this famous photograph is supernova 1987A, photographed with the AAT (Anglo-Australian Telescope) 2 months before it reached its maximum brightness. Superimposed on that image is a negative photograph of the region around the supernova copied from an AAT plate that was exposed in 1985, 2 years before the supernova was seen to explode. The precursor star appears to be a peculiar shape only because its image is blended with those of two other stars that happen to lie in the same line of sight. The brightest of the three stars had, in fact, exploded, and that was a star that had been previously observed and catalogued, as Sanduleak −69° 202. Image: © 1987 Anglo-Australian Observatory. Photograph by David Malin. (see color plate)

the case of the neutrinos detected from SN 1987A, at 07:36 UT on February 23, 1987, nine neutrinos were initially detected by Kamiokande within a 2-second window, followed by a further three between 9 and 13 seconds later. At the same time, the IMB detector bagged eight neutrinos in a 6-second time span, and Baksan registered five neutrino interactions in a 5-second span. The simultaneous detection of twenty-five neutrinos across three sites on Earth was an unprecedented and phenomenal achievement and remarkable proof that the theory of supernova collapse is well understood. Remember, neutrinos are virtually massless and they have no electrical charge. A neutrino will typically sail through the entire Earth as if it was not there because normal matter is surprisingly empty. To a neutrino, the earth looks a bit like a sphere of glass would to a photon. But when countless trillions are released in a supernova explosion, there are still enough for a few to interact with a 16-m-diameter vat of water 170,000 years later!

How many neutrinos are released in a core-collapse supernova explosion? Well, estimates vary, but 10^{57} is a popular figure. Yes, that is 10 to the power of 57 or 1 followed by 57 zeroes. Maybe you would prefer it in words? Okay, it is a billion trillion trillion trillion trillion neutrinos! Now, I'd like to indulge in a bit of speculative math. At a distance of 170,000 light-years, what fraction of those 10^{57} neutrinos will pass through a 16-m tank? Well 170,000 light-years is roughly 1.6×10^{21} m, and a sphere of that radius, through which many of the neutrinos will pass, 170,000 years after the core collapse, will have an area of 4π $(1.6 \times 10^{21})^2$ or 3.22×10^{43} square meters. Thus, one would expect $10^{57}/3.22 \times 10^{43} = 3.1 \times 10^{13}$ neutrinos to pass through each square meter (perpendicular to the supernova's direction) on Earth and, say, 256 times that number to pass through a 16-m-wide area. Of course, before reaching the tank, the neutrinos will have passed through the obligatory kilometer or more of rock before they get to the water-filled tank and, in doing so, some will have interacted with rock nuclei. However, I just wanted to give an idea of the neutrino flux passing through such a small area, 170,000 light-years away and 170,000 years later, to cause a few detections. One can see that, with only 12 neutrinos in the Kamiokande tank being detected from such a relatively nearby supernova, if the supernova had even been as far away as the Andromeda Galaxy (just over 2 million light-years) no neutrinos would have been detected. Conversely, if the supernova had been as close (3,000 light-years) as the one in Lupus, in AD 1006, one might have expected tens of thousands of neutrinos to be detected in each tank. The timing of the neutrinos' arrival at the respective tanks on February 23, 1987, is interesting. The detectors registered them at 07:36 UT. At 09:22 UT, almost 2 hours later, the veteran variable star observer (and comet discoverer) Albert Jones (based in New Zealand) was observing the Tarantula nebula in the Large Magellanic Cloud and did not spot the supernova visually through a small telescope. He later estimated it must have been fainter than mag 7.5. However, just over an hour later, Robert McNaught, at Coonabarabran, Australia, photographed the Large Magellanic Cloud and it turned out to be visible on his negatives, developed some time later. The actual discovery was not made until almost a day later when astronomer Ian Shelton, a University of Toronto research assistant working at the university's Las Campanas station, spotted the supernova on his 3-hour photograph at around 05:40 UT on February 24. The observatory night assistant, Oscar Duhalde, suspected it visually at 03:00 UT at about the same time; at 07:55 UT Duhalde estimated its magnitude as 4.5. Theory predicts that the core-collapse neutrinos will be emitted a few hours before the supernova erupts in the optical range, so Albert Jones' negative observation of February 23 is a valuable confirmation of the theory, especially as Jones' reputation as a reliable and accurate observer is second to none. Just over 3 hours after Shelton's discovery of the following day, Jones independently discovered SN 1987A in the finder of his telescope. Through drifting cloud, he estimated the magnitude as between 5.6 and 7.0. By 10:55 UT on that day (February 24), he was able to make an accurate estimate of mag 5.1. Shelton, Duhalde, and Jones were credited with the discovery, with McNaught providing the vital astrometric position of this historic find.

Unfortunately, due to the *Challenger* space shuttle disaster of 1986, the Hubble Space Telescope (HST) was not in orbit, in February 1987 in time to see SN 1987A go bang. Also, as most astronomers will recall, when it finally made it into orbit, the mirror was found to be defective. Southern Hemisphere observatories were able to study the region well though, and David Malin, at the Anglo-Australian Observatory, photographed the light-echoes as the "flash" from the explosion lit up the surrounding dust in the months after the explosion (see Figure 3.10). However, in recent years HST has regularly studied the expanding debris ring from the supernova (see

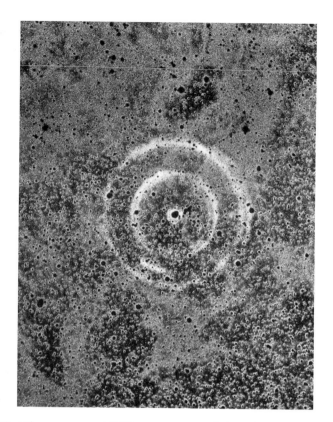

Figure 3.10. When supernova 1987A was seen to explode in the Large Magellanic Cloud, the brilliant flash of light had taken about 170,000 years to reach us. Some light was deflected by two sheets of dust near the supernova and is seen after the star has faded away because it traveled a fractionally longer distance to reach us. The dust responsible for the rings seen here lies in two distinct sheets, about 470 and 1,300 light-years from the supernova, and between us and the supernova. This picture (reproduced in the color plates section), made by subtracting images on plates taken before and after the supernova, is an accurate reproduction of the color of the extremely faint light echo, which in turn reflects the yellow color of the supernova when it was at its brightest, in May 1987. Image: © 1987 Anglo-Australian Observatory. Photograph by David Malin. (see color plate)

Figures 3.11, 3.12, and 3.13). At first, HST observed with the defective optics, but, after December 1993, the corrected optics (after the space shuttle repair mission) enabled HST's full resolution. The HST primary mirror has a diameter of 2.4 m, yielding a diffraction limited resolution of roughly 1/20 of an arc-second. At a distance of 170,000 light-years, this corresponds with almost 400 billion km, or 1/25 of a light-year, or 2 light-weeks. So, in theory, Hubble can resolve light-speed changes at that distance in a period as short as 2 weeks. Of course, the only changes that happen at that speed are going to be caused by light from the explosion traveling out and illuminating debris in the nearby environment 170,000 years ago.

So when the corrected optics were trained on the area of SN 1987A, what did they see? False-color images produced in August 1990, at the wavelength of doubly ionized oxygen, displayed the debris of the shattered star as a distorted pinky blob in the middle of a green/yellow ellipse. The ellipse represents material at a distance of roughly two-thirds of a light-year from the supernova but is *not* material from

Figure 3.11. This Hubble Space Telescope image taken in 1994, 7 years after the supernova 1987A exploded, is a medium-resolution shot of the region, after the blast from the supernova had long since faded away. The supernova resion is just above the centre of the image. Details below an arc-second can be glimpsed as well as the tiny hourglass figure shown in the next figure. Image: Hubble Heritage Team/AURA/STSCI/NASA.

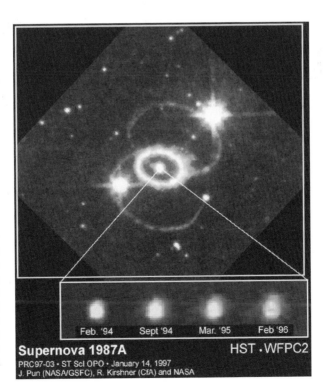

Figure 3.12. This ultra-high-resolution Hubble image of the 1987A supernova region was taken in January 1997 and shows features as small as a few light-weeks across. It also shows the mysterious hourglass feature described in the text. Image: NASA (J. Pun)/GSFC/CfA (R. Kirshner).

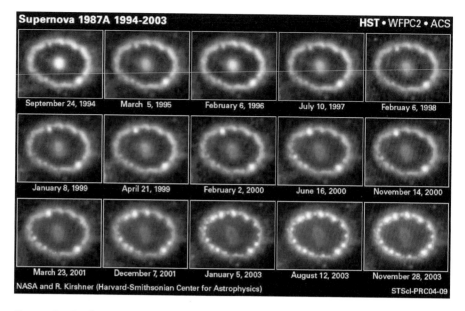

Supernova 1987A 1994-2003 **HST • WFPC2 • ACS**

September 24, 1994 | March 5, 1995 | February 6, 1996 | July 10, 1997 | Februay 6, 1998

January 8, 1999 | April 21, 1999 | February 2, 2000 | June 16, 2000 | November 14, 2000

March 23, 2001 | December 7, 2001 | January 5, 2003 | August 12, 2003 | November 28, 2003

NASA and R. Kirshner (Harvard-Smithsonian Center for Astrophysics) STScI-PRC04-09

Figure 3.13. These 15 images show the changes in the region immediately surrounding SN 1987A from 1994 to 2003. The ellipse represents material at a distance of roughly two-thirds of a light-year from the supernova. Image: NASA and R. Kirshner (Harvard-Smithsonian Center for Astrophysics). (see color plate)

the supernova explosion. It is thought to be a ring of debris, possibly expelled from the blue supergiant progenitor star, which existed long before the star collapsed. A flash of ultraviolet light from the supernova probably excited the old ring when the light arrived at it some 8 months after the explosion. The ring may well be the "waist" of an hourglass-like feature. In February 1994, Hubble took a very deep image of the 1987A field at hydrogen-alpha wavelengths and confirmed two more rings that had only been suspected, but not proved, in other images. The two new rings are twice the size of the bright inner ring and not concentric with that ring. Their nature has prompted much debate. They appear to be between 2 and 3 light-years from SN 1987A and may be caused by beams of high-energy particles "writing" an emission pattern on interstellar material (like a laser beam at a rock concert). This might be possible if the supernova progenitor star had a binary companion, such as a black hole or neutron star, with its own nearby companion. Matter falling from the companion onto the superdense object would be heated and blasted back into space along two very narrow jets, along with an accompanying beam of radiation. There seems to be far more to the SN 1987A story than just one star going bang. As with so many violent phenomena observed by ground-breaking instrumentation, at first there are more new questions than there are answers to old ones. Supernovae are rarely identical. They do not come off a production line. By 1997, 10 years after we saw SN 1987A explode, Hubble revealed evidence of the first collision between the actual blast wave and the inner bright ring surrounding the former star Sanduleak −69° 202. In those same images, the visible part of the debris cloud was revealed as a mainly sickle-shaped feature, a matter of light-weeks from the remains of the star (i.e., at the very resolution limit of the Hubble Space Telescope).

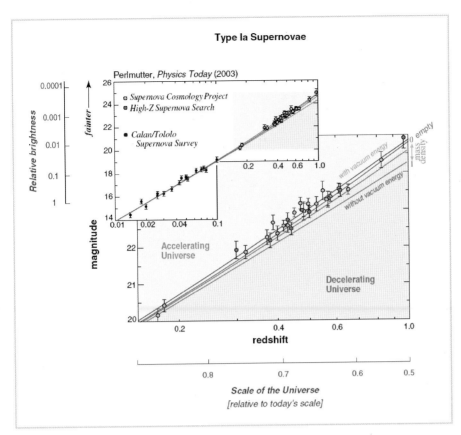

Figure 2.4. Observed magnitude versus redshift for well-measured distant and (in the inset) nearby Type Ia supernovae. The older Calan/Tololo Supernova Survey results merge with the more recent and unexpected results on distant supernovae made by the Supernova Cosmology Project and the High-z Supernova Search Team. Diagram: By kind permission of Saul Perlmutter (originally reproduced in *Physics Today*, 2003).

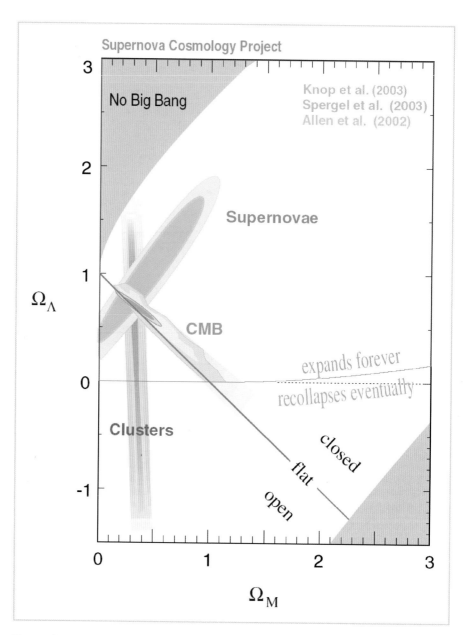

Figure 2.5.

Figure 2.5. Confidence regions for Omega Mass versus Omega Lambda. This diagram shows the confidence in the cosmological models for the kind of Universe we live in. The models of the 1980s and 1990s have been totally shattered by the distant supernova findings. If we had all the information (e.g., from an all-powerful and knowledgeable being), we could stick a pin on a specific point on the graph and say "this is how our universe is." The x axis represents the relative mass density of the Universe, Omega M (i.e., how much gravity producing potential there is). Obviously, the dark matter content is of crucial importance here. The y axis represents the acceleration parameter, or rather the Cosmological Constant's relative density, or Omega Lambda, which is now associated with the term dark energy and the accelerated expansion of the Universe. Note, this is Omega Lambda, not the Cosmological Constant itself, but the ratio of the "density" of the cosmological constant to the critical zero point. In a theoretical Universe where, eventually, the Big Bang expansion will be exactly halted by the gravitational influence of all the matter, the Omega Lambda value on this graph would be zero. The critical supernova ellipses in the mid/top left of the diagram show the Universe model confidence contours based on ultradistant Type Ia supernova redshift and brightness measurements. These contours illustrate the amazing result that the Universe is still accelerating (and faster than in the past). At the time of writing, a y-axis acceleration parameter of around 0.7 and an x-axis relative mass density of around 0.3 seem most likely, especially when combined with Cosmic Microwave Background measurements (CMB on the graph) and measurements of the dark matter content in galaxy clusters (Clusters on the graph). Essentially, this means that the Universe started with a Big Bang, will expand forever, and, also, is essentially flat. By flat we mean that the fabric of space obeys Euclidean geometry and that a small chunk of the whole cosmos expanded very rapidly in the first instant after the Big Bang to form the almost perfectly smooth background radiation of the observable Universe we see today. Mathematically, a flat Universe is one in which the relative density parameters Omega M plus Omega Lambda equal one. If there were no need for a Cosmological Constant (as cosmologists imagined less than 10 years ago), this flat point would occur at the point where Omega M was unity and Omega Lambda was zero. The strongest evidence for flatness comes from WMAP (the Wilkinson Microwave Anisotropy Probe) CMB measurements, and the evidence that our Universe is flat constrains our Universe to the diagonal line in Figure 2.5. Thus, on current evidence our Universe fits within the lower part of the supernova confidence ellipse where it is crossed by the diagonal line marked flat and where the CMB and Clusters lines also intersect. Prior to the relatively recent supernova evidence that the acceleration of the Universe was increasing, this would all have been considered extremely unlikely, if not preposterous! Because the Universe is thought to be flat, and so Omega M plus Omega Lambda, by definition, equals one, or 100%, the Universe is now often described in the media as consisting of 70% dark energy and 30% (dark matter + visible matter). Diagram: By kind permission of Saul Perlmutter, et al., The Supernova Cosmology Project, based on Knop et al. (2003).

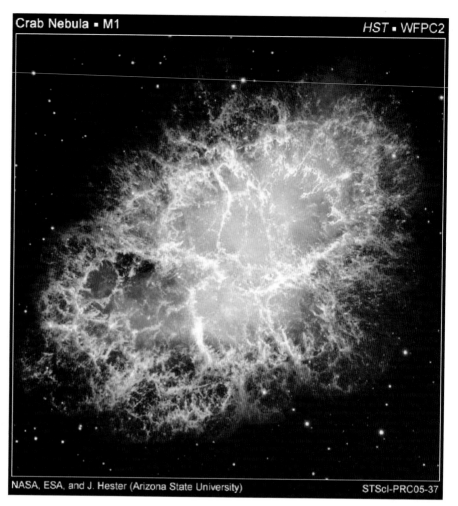

Crab Nebula ▪ M1

HST ▪ WFPC2

NASA, ESA, and J. Hester (Arizona State University)

STScI-PRC05-37

Figure 3.1. The Crab Nebula imaged by the Hubble Space Telescope in 1999 and 2000. This amazing image is a mosaic of 24 individual images. Image: NASA, ESA, J. Hester and A. Loll (Arizona State University).

Figure 3.7. The very bright star in the bottom right of this image is the supernova 1987A in the Large Magellanic Cloud (LMC): the first supernova to be visible to the naked eye for 383 years! The complex red feature in the top left is the LMC's famous Tarantula nebula. The angular width of this image is about 28 arc-minutes. Image: © 1987 Anglo-Australian Observatory. Photograph by David Malin.

Figure 3.8. The picture shows the field of supernova 1987A in the Large Magellanic Cloud 10 days after the explosion and (on the right) before the explosion. The original progenitor star, Sanduleak −69° 202, can be seen on the right-hand image. Image: © 1987 Anglo-Australian Observatory. Photograph by David Malin.

Figure 3.9. The brightest star in this famous photograph is supernova 1987A, photographed with the AAT (Anglo-Australian Telescope) 2 months before it reached its maximum brightness. Superimposed on that image is a negative photograph of the region around the supernova copied from an AAT plate that was exposed in 1985, 2 years before the supernova was seen to explode. The precursor star appears to be a peculiar shape only because its image is blended with those of two other stars that happen to lie in the same line of sight. The brightest of the three stars had, in fact, exploded, and that was a star that had been previously observed and catalogued, as Sanduleak −69° 202. Image: © 1987 Anglo-Australian Observatory. Photograph by David Malin.

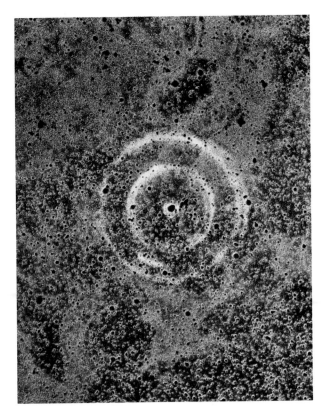

Figure 3.10. When supernova 1987A was seen to explode in the Large Magellanic Cloud, the brilliant flash of light had taken about 170,000 years to reach us. Some light was deflected by two sheets of dust near the supernova and is seen after the star has faded away because it traveled a fractionally longer distance to reach us. The dust responsible for the rings seen here lies in two distinct sheets, about 470 and 1,300 light-years from the supernova, and between us and the supernova. This picture made by subtracting images on plates taken before and after the supernova, is an accurate reproduction of the color of the extremely faint light echo, which in turn reflects the yellow color of the supernova when it was at its brightest, in May 1987. Image: © 1987 Anglo-Australian Observatory. Photograph by David Malin.

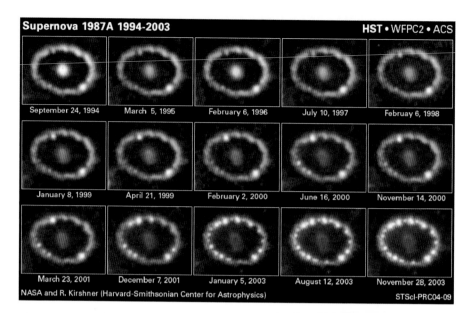

Supernova 1987A 1994-2003   **HST • WFPC2 • ACS**

September 24, 1994	March 5, 1995	February 6, 1996	July 10, 1997	Februay 6, 1998
January 8, 1999	April 21, 1999	February 2, 2000	June 16, 2000	November 14, 2000
March 23, 2001	December 7, 2001	January 5, 2003	August 12, 2003	November 28, 2003

NASA and R. Kirshner (Harvard-Smithsonian Center for Astrophysics)   STScI-PRC04-09

Figure 3.13. These 15 images show the changes in the region immediately surrounding SN 1987A from 1994 to 2003. The ellipse represents material at a distance of roughly two-thirds of a light-year from the supernova. Image: NASA and R. Kirshner (Harvard-Smithsonian Center for Astrophysics).

Figure 13.7. The giant Vela Supernova Remnant. This image is 2 degrees wide. Image: © 1987 Anglo-Australian Observatory. Photograph by David Malin.

Chapter 4

The Top 100 Extragalactic Supernovae

Despite more than 300 supernova discoveries per year in recent years, really bright ones, in external galaxies, easily visible in medium- or large-sized amateur telescopes, are a relatively rare sight. However, the top 100 have all been within that category. Although I am excluding the superbright naked-eye Milky Way supernovae of historical times in this top 100 listing of Table 4.1, I feel the supernova 1987A in the Large Magellanic Cloud must be included. First, it is definitely outside the Milky Way, and, second, it is so relatively recent that many Southern Hemisphere readers of this book will remember seeing it. The vast bulk of supernovae have been discovered since the start of the 20th century; namely, as soon as routine photographic patrolling started.

The Brightest of All

In the past 120 years, only 13 extragalactic supernova discoveries of magnitude 11.5 and brighter were made, and many more must have been missed in the early years of that period. As soon as we include supernovae fainter than magnitude 12.0, the number of discoveries soars. In Table 4.1, after the 13 brightest supernovae, there are 30 more between magnitude 12.0 and 12.9 Also, especially in the early days, magnitude estimates where not photometrically precise, and the quoted peak magnitude and discovery magnitude may occasionally have been interchanged. This means there is some uncertainty in defining which supernovae were, say, the 10th, 11th, or 12th brightest. In general, supernovae peak in brightness within a week of detonation/discovery, and the table, constructed mainly from the IAU (International Astronomical Union) circular magnitudes, is primarily based on the discovery mag. However, there are exceptions to this, and the supernova 2002ap in M 74, which eventually peaked at around magnitude 12.5, two full magnitudes brighter than its discovery magnitude, is a prime example. Sometimes supernovae are caught very early, on the rise, and then brighten by two or more magnitudes after discovery. The brightest extragalactic supernova, 1987A, also took many weeks to peak in brightness, at magnitude 2.8, after its discovery.

For me, as an amateur astronomer, the ultrabright supernovae of 12th mag and better are really something special, because, through a large amateur telescope, you

Table 4.1. The Top 100 Extragalactic Supernovae (1885–2005)

Desig	Galaxy	Mes/Cald	Date	RA	DEC	Offset	Mag	Am. Discoverers/Comments
1987A	LMC		1987 02 24	05 35.4	−69 16	4S	4.5	Peaked at mag 2.8
1885A	NGC 224	M31	1885 08 17	00 42.7	+41 16	15W 4S	5.8	
1895B	NGC 5253		1895 07 07	13 39.9	−31 39	16E 23N	8.0	
1937C	IC 4182		1937 08 16	13 05.8	+37 36	30E 40N	8.4	Well Observed Supernova
1972E	NGC 5253		1972 05 06	13 39.9	−31 39	38W 100S	8.5	
1954A	NGC 4214		1954 04 10	12 15.6	+36 20	84E 216S	9.8	Disc. a month past max.
1993J	NGC 3031	M81	1993 03 28	09 55.4	+69 01	45W 160S	10.2	Garcia
1921C	NGC 3184		1921 12 05	10 18.3	+41 25	79E 236S	11.0	
2004dj	NGC 2403	C7	2004 07 31	07 37.3	+65 36	160E 10N	11.2	
1961H	NGC 4564		1961 05 02	12 36.4	+11 26	0 5N	11.2	Itagaki
1980K	NGC 6946	C12	1980 10 28	20 34.9	+60 09	280E 166S	11.4	
1971I	NGC 5055	M63	1971 05 24	13 15.7	+42 01	2W 147S	11.5	
1970G	NGC 5457	M101	1970 07 30	14 03.3	+54 21	97W 370S	11.5	Peculiar
1960F	NGC 4496		1960 04 17	12 31.7	+03 56	38E 24N	11.6	
1962M	NGC 1313		1962 11 26	03 18.2	−66 29	0 150S	11.7	
1920A	NGC 2608		1920 01 01	08 35.2	+28 29	19W 5N	11.8	
1968L	NGC 5236	M83	1968 07 17	13 37.1	−29 52	5W 0	11.9	
1939B	NGC 4621	M59	1939 05 19	12 42.0	+11 39	0 53S	11.9	
1939A	NGC 4636		1939 01 17	12 42.9	+02 42	26W 20N	11.9	
1983U	NGC 3227		1983 11 04	10 23.6	+19 52	12W 0	12.0	
1960R	NGC 4382	M85	1960 12 20	12 25.4	+18 11	8E 132S	12.0	
1979C	NGC 4321	M100	1979 04 19	12 22.9	+15 49	56E 87S	12.1	
1909A	NGC 5457	M101	1909 01 26	14 03.3	+54 21	620W 408N	12.1	
1989M	NGC 4579	M58	1989 06 28	12 37.6	+11 49	40W 33N	12.2	
1961V	NGC 1058		1961 12 05	02 43.4	+37 21	76E 17N	12.2	
1981B	NGC 4536		1981 03 02	12 34.6	+02 11	36E 36N	12.3	

Table 4.1. The Top 100 Extragalactic Supernovae (1885–2005) (Continued)

Desig	Galaxy	Mes/Cald	Date			RA		DEC		Offset		Mag	Am. Discoverers/Comments
1956A	NGC 3992	M109	1956	03	08	11	57.6	+53	22	67E	9S	12.3	
1954B	NGC 5668		1954	04	27	14	33.4	+04	27	2W	20S	12.3	
1919A	NGC 4486	M87	1919	02	24	12	30.8	+12	23	15W	100N	12.3	
2003hv	NGC 1201		2003	09	09	03	04.2	−26	05	17E	57S	12.5	
1986G	NGC 5128	C77	1986	05	03	13	25.6	−43	02	120E	60S	12.5	Evanspeculiar
1985L	NGC 5033		1985	06	13	13	13.5	+36	36	68W	51N	12.5	
1983N	NGC 5236	M83	1983	07	03	13	37.1	−29	52	120W	130S	12.5	Evans
1980N	NGC 1316		1980	12	07	03	22.6	−37	14	220E	20S	12.5	
1957B	NGC 4374	M84	1957	04	23	12	25.0	+12	53	8W	47N	12.5	
1935C	NGC 1511		1935	08	16	03	59.5	−67	38	55E	8S	12.5	
1895A	NGC 4424		1895	03	16	12	27.2	+09	25	75E	11S	12.5	
2002ap	NGC 628	M74	2002	01	29	01	36.4	+15	45	258W	108S	12.5	Peak Mag. Hirose Hypernova
1981D	NGC 1316		1981	03	01	03	22.6	−37	14	20W	100S	12.7	Evans
2005af	NGC 4945	C83	2005	02	08	13	04.7	−49	34	407W	351S	12.8	
2004et	NGC 6946	C12	2004	09	27	20	35.4	+60	07			12.8	
1992A	NGC 1380		1992	01	11	03	36.4	−34	57	3W	62N	12.8	
1985S	MCG 020710		1985	09	19	02	27.5	−10	10	10E	10S	12.8	
1971L	NGC 6384		1971	06	24	17	32.4	+07	04	27E	20N	12.8	
1969L	NGC 1058		1969	12	02	02	43.4	+37	21	190E	110S	12.8	
1940B	NGC 4725		1940	05	05	12	50.4	+25	30	95E	118N	12.8	
1937D	NGC 1003		1937	09	09	02	39.3	+40	53	48E	1S	12.8	
1978G	IC 5201		1978	11	24	22	21.4	−46	04	96W	42N	12.9	
1964E	UGC 6983		1964	03	12	11	59.2	+52	42	83W	44S	12.9	
1995al	NGC 3021		1995	11	01	09	50.9	+33	33	15W	3S	13.0	
1974G	NGC 4414		1974	04	20	12	26.5	+31	13	27E	56S	13.0	
1970L	NGC 2968		1970	10	31	09	43.2	+31	56	120E	75N	13.0	
1969Q	NGC 4472	M49	1969	06		12	29.7	+08	00	120E	0	13.0	
1967C	NGC 3389		1967	02	28	10	48.4	+12	32	43W	44N	13.0	
1966J	NGC 3198		1966	12	20	10	20.0	+45	34	100W	165S	13.0	
1962H	IC 4237		1962	06	28	13	24.5	−21	08	19W	3S	13.0	

**The Top 100
Extragalactic
Supernovae**

Table 4.1. The Top 100 Extragalactic Supernovae (1885–2005) (Continued)

Desig	Galaxy	Mes/Cald	Date	RA	DEC	Offset	Mag	Am. Discoverers/Comments
1961I	NGC 4303	M61	1961 06 03	12 21.9	+04 28	82E 12S	13.0	
1939C	NGC 6946	C12	1939 07 17	20 34.9	+60 09	215W 24N	13.0	
1912A	NGC 2841		1912 02 19	09 21.9	+50 59	50W 20N	13.0	
1961F	NGC 3003		1961 02 21	09 48.6	+33 25	34E 17N	13.1	
2003gd	NGC 628	M74	2003 06 12	01 36.7	+15 44	13E 16S	13.2	Evans
1941A	NGC 4559	C36	1941 02 05	12 35.9	+27 57	30W 26N	13.2	
2004gk	IC 3311		2004 11 25	12 25.6	+12 16	2E 3N	13.3	
1964L	NGC 3938		1964 12 11	11 52.8	+44 06	3W 31N	13.3	
1959D	NGC 7331	C30	1959 06 28	22 37.1	+34 26	32W 13N	13.4	
1999em	NGC 1637		1999 10 29	04 41.5	-02 52	15W 17S	13.5	
1997X	NGC 4691		1997 02 01	12 48.2	-03 20	7E 0N	13.5	Aoki
1994W	NGC 4041		1994 07 29	12 02.2	+62 08	9W 18N	13.5	
1994I	NGC 5194	M51	1994 04 02	13 29.9	+47 12	14E 12S	13.5	Puckett et al.
1992ad	NGC 4411		1992 07 01	12 26.8	+08 52	40E 40S	13.5	Evans
1990M	NGC 5493		1990 06 15	14 11.5	-05 03	15W 4N	13.5	Evans
1988A	NGC 4579	M58	1988 01 16	12 37.6	+11 49	0 40S	13.5	Ikeya, Evans et al.
1986L	NGC 1559		1986 10 07	04 17.6	-62 48	50W 0	13.5	Evans
1984J	NGC 1559		1984 07 27	04 17.6	-62 48	25W 10S	13.5	Evans
1983V	NGC 1365		1983 11 25	03 33.6	-36 08	57W 30S	13.5	Evans
1983I	NGC 4051		1983 05 11	12 03.2	+44 31	21E 50S	13.5	
1981A	NGC 1532		1981 02 24	04 12.1	-32 52	70E 15N	13.5	
1980L	NGC 7448		1980 10 08	23 00.1	+15 58	27W 4N	13.5	Evans
1979B	NGC 3913		1979 02 28	11 50.6	+55 20	40E 20N	13.5	
1968I	NGC 4981		1968 04 23	13 08.7	-06 47	2E 7N	13.5	

Table 4.1. The Top 100 Extragalactic Supernovae (1885–2005) (Continued)

Desig	Galaxy	Mes/Cald	Date	RA	DEC	Offset	Mag	Am. Discoverers/Comments
1968D	NGC 6946	C12	1968 02 29	20 34.9	+60 09	45E 20N	13.5	
1965I	NGC 4753		1965 06 18	12 52.4	−01 12	98W 68N	13.5	
1937F	NGC 3184		1937 12 09	10 18.3	+41 25	5E 149S	13.5	
1921B	NGC 3184		1921 04 06	10 18.3	+41 25	32E 160S	13.5	
1907A	NGC 4674		1907 05 09	12 46.0	−08 39	10W 11N	13.5	
2005cs	NGC 5194	M51	2005 06 28	13 29.9	+47 10	67S	13.5	
1996X	NGC 5061		1996 04 12	13 18.0	−26 51	52W 31S	~13	Evans; Takamizawa
1998bu	NGC 3368	M96	1998 05 09	10 46.8	+11 50	4E 55N	~13	Villi
1991T	NGC 4527		1991 04 13	12 34.2	+02 39	26E 44N	~13	Evans; Villi et al., Peculiar
1989B	NGC 3627	M66	1989 01 30	11 20.2	+13 01	15W 50N	~13	Evans
1985B	NGC 4045		1985 01 17	12 02.8	+01 58		~13	
1984R	NGC 3675		1984 12 02	11 26.1	+43 35		~13	
1983P	NGC 5746		1983 07 11	14 44.8	+01 57	7E 5S	~13	Okazaki; Evans
1983G	NGC 4753		1983 04 04	12 52.4	−01 12	10W 20S	~13	
1980O	NGC 1255		1980 10 30	03 13.6	−25 47	73W 59S	~13	
1971R	IC 4798		1971 10 01	18 58.2	−62 06	12S 0	~13	
1971G	NGC 4165		1971 03 29	12 12.2	+13 14	3E 30S	13.6	
1962J	NGC 6835		1962 08 28	19 54.6	−12 34	42W 23S	13.6	
1959C	UGC 8263		1959 06 28	13 11.3	+03 24	7E 3S	13.6	
2005df	NGC 1559		2005 08 04	04 17.6	−62 46	15E 40N	13.7	EVANS. Peak mag 12.5

The table above shows the 100 brightest extragalactic supernovae, listed primarily by discovery magnitude with their designation, host galaxy, Messier or Caldwell number of the galaxy, discovery date, right ascension and declination, offset in arc-seconds from the nucleus, discovery magnitude, and the principal amateur astronomers named. Much of the information in this table has been extracted from the CBAT/IAU Web page at http://cfa-www.harvard.edu/iau/cbat.html. Supernovae that were definitely spectroscopically confirmed as Type Ia (or from their light curve) appear in a darker gray font. Supernovae 1971I, 1986G, and 1991T were especially peculiar Type Ia's. SN 2002ap (see Figure 4.1) was initially discovered at magnitude 14.5 but peaked at magnitude 12.5. It was an unusual hypernova that slowly brightened to its peak brightness. Note the 16 discoveries by Evans in the top 100, half of them prestigious Type Ia's.

Figure 4.1. Supernova 2002ap was a very bright "hypernova" discovered by Hirose in Japan. Although it was discovered at magnitude 14.5, it slowly rose to magnitude 12.5, becoming one of the brightest ever supernovae. Image: M. Mobberley.

can often stare at them with direct vision; there is none of this averted vision hassle, where you look to one side of the object so the most sensitive part of the retina detects the star (see Chapter 6). I was fortunate to have just started using a 0.49-m Newtonian when SN 1993J peaked at magnitude 10.5 (see Table 4.1); a very memorable view indeed. That supernova, in Messier 81, was the brightest one of the amateur CCD (Charge Coupled Device) era and came along just as many astrophotographers were switching to electronic imaging (see Figure 4.2).

Supernova 1980K, in NGC 6946 (no. 11 in the table), was well observed by many U.K. observers and was the sixth supernova this most-productive galaxy has produced (see Figure 4.3). The galaxy produced a seventh in 2002 and an eighth in 2004. Visual observers, with a variety of telescopes, observed SN 1980k as it dropped from 11th to 15th magnitude, after which point the photographers took over.

The ninth object in the list, 2004dj in NGC 2403, was a very recent object. The discovery, by Itagaki, was made when the galaxy was very badly placed in the 2004 Northern Hemisphere summer sky. But it was still an easy CCD object in the following winter months (see Figures 4.4 and 4.5).

The other supernovae in the list below are rather further back in history, but just look at NGC 5253, third and fifth in the rankings. What a galaxy! It produced two 8th magnitude supernovae in the space of 77 years. Wow! What a pity for U.K. and U.S. observers that it resides at the southerly declination of −31 degrees.

A quick scan down the top 100 table reveals a large number of Messier and Caldwell designations amongst the, predominantly, NGC objects; 32 discoveries in fact. As most amateur astronomers will know, the Messier catalogue, devised by Charles Messier in the 18th century, predates the NGC catalogue and was mainly compiled as an aide to prevent Messier thinking these "fuzzy objects" were new comets. Not surprisingly, the 39 (or 40 if you really believe NGC 5866 is justified in being M 102) Messier galaxies are, with a few exceptions, the brightest ones that can be seen from the latitude of Paris, where Messier lived. Conversely, the Caldwell catalogue is a very modern affair and its inventor, the famous amateur astronomer and author Patrick Moore (Patrick Caldwell-Moore to be precise), did *not* discover the galaxies. However, he did have a very good idea, namely, to make a catalogue to

Figure 4.2. Supernova 1993J in M 81, discovered by Francisco Garcia Diaz, was the brightest Northern Hemisphere supernova for more than 30 years. These two images, by the author, show its fade from 10th to 16th magnitude between April 1993 and January 1994.

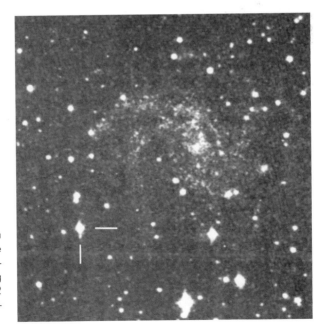

Figure 4.3. Supernova 1980k in the productive galaxy NGC 6946. Photograph by Brian Manning using a 260-mm f/7.2 Newtonian and hypersensitized Kodak 2415 film.

Figure 4.4. The bright supernova 2004dj in NGC 2403. Image by Jeremy Shears with a Takahashi 102-mm refractor.

Figure 4.5. The same supernova, 2004dj in NGC 2403, as imaged by the Hubble Space Telescope. Image: NASA.

include the easy (with modern amateur telescopes) and impressive galaxies that Messier did not include in his catalogue and extend this into the Southern Hemisphere skies. When you add the Messier and Caldwell galaxies together, you have 74 objects that are easily visible visually through an amateur telescope.

In my table of the top 100 supernovae, there are 23 Messier galaxy supernovae and 9 Caldwell galaxy supernovae. But M 51, M 58, M 74, and M 101 have each produced two supernovae in the top 100, and Caldwell 12 (the ultraproductive NGC 6946) has single-handedly produced four supernovae in the top 100. Thus, we have 19 Messier and 6 Caldwell galaxies producing supernovae in the top 100. I will have more to say about patrolling the Messier and Caldwell galaxies in Chapters 11 and 12.

Science from the Brightest Supernovae

We have already seen that supernova 1987A in the Large Magellanic Cloud provided mountains of scientific data for professional astronomers, not least from the detected neutrinos. It was exceptionally fortunate that such a nearby supernova should detonate just as an array of high-technology equipment was in place to analyze such an event. But SN 1987A was not the only bright supernova to provide masses of scientific data. Six years later, on Sunday, March 28, 1993, Spanish amateur astronomer Francisco Garcia Diaz from Lugo discovered a new object in the field of the Messier galaxy M 81, also known as NGC 3031. In 1993, amateur supernova discoveries were rare, unless, that is, your name was the Rev. Robert Evans. By March 1993, that amazing Australian had already discovered 27 supernovae visually, but no other amateur was in his league, and automated amateur CCD patrols were in the future. However, Garcia did succeed and M 81 is much too far north for Evans to patrol. Garcia spotted supernova 1993J visually, with his 25-cm f/3.9 Newtonian telescope, at a magnification of 111×, as an 11th magnitude star. It would later reach a maximum brightness of about mag 10.5, around March 31. M 81 is one of the largest galaxies visible in the night sky, mainly because it is so relatively nearby. The best estimates place it at a distance of only 12 million light-years, or about 70 times further than SN 1987A in the Large Magellanic Cloud. So while SN 1993J was never going to be a naked-eye object, it was certainly a treat for Northern Hemisphere telescope users who missed out when SN 1987A erupted 6 years earlier. But like 1987A, there were mysteries over this new supernova. SN 1993J started out as a classic Type II supernova, with hydrogen lines in its spectrum. However, within weeks these had become less prominent and helium lines appeared, making SN 1993J appear more like a Type Ib supernova with a dense helium core. In addition, after the initial rise to a peak brightness of mag 10.5, which took a few days, SN 1993J dimmed for a week, to mag 11.8 and then exhibited a second maximum 2 weeks later. It then declined steadily at an average rate of roughly 1 mag every 50 days. This bizarre performance does not match the typical light curve of Type II-L, Type II-P, or even Type II-n supernovae. The best explanation for both the double-peaked light curve and the apparent hydrogen deficiency is that maybe the original star had a much smaller hydrogen envelope than normal: possibly just a fraction of a solar mass, instead of many solar masses. (Note: a solar mass means a mass equivalent to our own sun.) A hydrogen deficiency like this might produce an initial bright peak in the light curve and would certainly explain the hydrogen-deficient spectra. What might cause such a hydrogen deficiency? Well, stellar winds from the blast might have carried the hydrogen away, but a more likely explanation is that a companion star might have gravitationally stolen a large proportion of the supernova star's hydrogen. SN 1993J is one of only a few stars for which a progenitor has been identified (i.e., pictures of the 20th magnitude star, before it exploded, exist). This is hardly surprising when you consider that a supernova is around 10 magnitudes fainter before it goes bang. For galaxies farther away than 10 million light-years (i.e., the overwhelming majority), this implies that a progenitor star will be fainter than magnitude 20. Only now, in the era of Hubble and massive telescopes like the Keck and the VLT (Very Large Telescope in Cerro Paranal, Chile), are such deep galaxy masters available. But even if ultradeep images can be located, the fields are often so cluttered that it is by no means obvious which star was the progenitor, until many years later, when the initial explosion has faded and a "missing star" can be found. If the star was in a binary system, the situation can be very complicated. Supernovae 1987A, 1993J, and 2005cs (see Figures 4.6 and 4.7) all have very good progenitor candidates, and the progenitor of SN 1961V in NGC 1058 was almost (but not quite) pinned down in the photographic era.

Although this book is primarily about what we currently know about supernovae and how to observe them, it would be remiss not to mention a few of the very brightest historical extragalactic examples. M 31 in Andromeda is undoubtedly the best known naked-eye galaxy. Unlike the Magellanic Clouds, it does actually look like a proper spiral galaxy, even with the oblique viewing angle. On August 20, 1885, Ernst Hartwig (1851–1923) observing at the Dorpat Observatory in Estonia discovered a new star in M 31. It had achieved magnitude 5.8 (probably) on August 17 and, despite the fact that it was independently discovered by several other observers, Hartwig was the only discoverer to really appreciate the object's importance. It was designated as the variable star S Andromedae. Within 5 years, it had faded to magnitude 16. More than 100 years after its discovery, the remnant of SN 1885 was pinned down by R.A. Fessen and his colleagues using the 4-m Mayall telescope at Kitt Peak Observatory in Arizona, USA. Observations by the

Figure 4.6. The bright supernova 2005cs in Messier 51, imaged by the author.

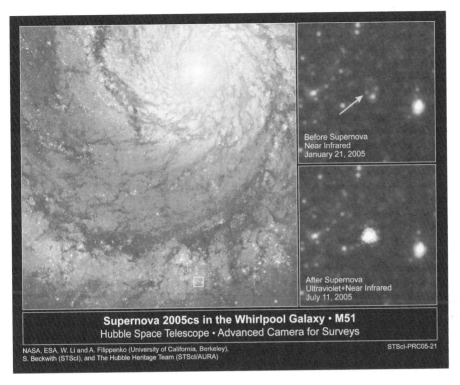

Before Supernova
Near Infrared
January 21, 2005

After Supernova
Ultraviolet+Near Infrared
July 11, 2005

Supernova 2005cs in the Whirlpool Galaxy • M51
Hubble Space Telescope • Advanced Camera for Surveys

NASA, ESA, W. Li and A. Filippenko (University of California, Berkeley),
S. Beckwith (STScI), and The Hubble Heritage Team (STScI/AURA)

STScI-PRC05-21

Figure 4.7. The progenitor star for SN 2005cs in M 51, before it exploded, as detected in high-resolution Hubble Space Telescope images. Image: NASA, ESA, Weidong Li and Alex Filippenko (University of California, Berkeley), S. Beckwith (STSCI), and the Hubble Heritage Team (STSCI/AURA).

Hubble Space Telescope in 1999 provided evidence for the remnant containing between 0.1 and 1.0 solar masses of iron.

Ten years later, in 1895, the next two extragalactic supernovae were discovered. The first, in NGC 4424, was discovered on March 16 that year by Wolf and designated as VW Vir. The second was a superb 8th magnitude discovery in the Southern Hemisphere galaxy (declination −31° 39′) NGC 5253. This latter discovery, by Fleming, was given the variable star designation Z Centauri and was observed for more than 400 days after discovery, although some of the visual photometry was not particularly accurate. As we have already seen, NGC 5253 would produce another superb supernova 77 years later, in 1972. Yet again, this would be an 8th magnitude object. Although these 19th century supernova discoveries were very bright, the first really well observed supernovae did not flare up until much later; 1937 in fact. Supernova 1937C, in the less than impressive galaxy IC 4182, was discovered by the indefatigable Fritz Zwicky on August 16 of that year, at magnitude 8.4; it was really the first Type Ia supernova to be observed accurately over a very long timescale. The supernova shone at three magnitudes (16 times) brighter than the brightness of the parent galaxy! Twenty-two reliable observations of 1937C were made more than 300 days after maximum brightness and this helped Minkowski, in 1940, to divide supernovae into two distinct types by virtue of their light curves. The light curves of 8th magnitude 1937C in IC 4182, derived from photographs, and of 8th magnitude 1972E in NGC 5253 appear to have been almost identical. It should be remembered that while such bright supernovae were easy to observe and photograph in the 1920s and 1930s, they were incredibly difficult to discover. Without automated patrol equipment, the task of supernova discovery was limited to tedious photographic patrols. Computerized automatic telescopes that could slew to different galaxies were decades in the future. Only 7 supernovae were discovered in the 1920s, and a mere 19 in the 1930s. The Second World War dropped the tally to 15 during the whole of the 1940s. But the 1950s saw a dramatic surge in supernova discovery with 122 discoveries in that decade, 74 of which were made by Fritz Zwicky who discovered a total of 123 supernovae in his lifetime. All but one of these were discovered between 1936 and 1973, but Zwicky's first super-nova discovery was made in 1921, making his discoveries span a colossal 52 years! So the task of analyzing supernova light curves and dividing them into different categories was rather dependent on discovering them in that early part of the 20th century. In addition, even when the supernovae were bright, they were frequently discovered after they had peaked in brightness, making the early part of the light curve a mystery. The number of supernovae discovered in recent years is colossal. In 2005, a record 366 supernovae were discovered (compared with 249 in 2004). Eighty of these were in the brighter NGC and IC catalogue galaxies. One hundred forty nine of those 366 became brighter than magnitude 18 and thus were poten-tial amateur CCD discovery targets. Amazingly, 17 extragalactic novae (mainly in the nearby galaxies M 33 and M 31) were also discovered in 2005.

One of the best observed supernovae in the past few years was also one of the brightest supernovae of all time. SN 2004dj in NGC 2403 became the ninth bright-est extragalactic supernova discovery when it was discovered by the Japanese supernova hunter Itagaki on July 31 of that year (based on his magnitude estimate of 11.2) At the time of discovery, NGC 2403 was fairly badly placed below the north celestial pole, but this did mean that the galaxy, also well-known as Caldwell 7, was superbly placed by the following winter when amateur astronomers with large tele-scopes or CCDs continued to follow its decline. (See Chapter 7 and the Figure 7.25 for more details).

Almost a quarter of a century earlier, the bright supernova 1980K, in the pro-ductive galaxy NGC 6946, produced a deluge of amateur photometric results as it faded from its discovery magnitude of 11.4. This supernova was a dream object for amateur astronomers as it was not only bright but situated over 5 arc-minutes from the galaxy center, so its brightness could not be confused with the background light from the galaxy.

Chapter 5

Supernovae: A Threat to Life on Earth

The prospect of a naked-eye supernova in our own galaxy is something that many astronomers, amateur and professional, hope they live to see. Of course, for many Southern Hemisphere astronomers, this dream was almost realized in 1987 when SN 1987A erupted in the Large Magellanic Cloud. However, although that object did reach an impressive magnitude 2.8 in May of that year, it was not quite the Venus rivaling object that astronomers had hoped for: at 170,000 light-years it was just too far away. There is no doubt that a true galactic supernova (i.e., one in our own Milky Way Galaxy) could be truly spectacular. We have already seen that the supernova of 1006 reached magnitude −9.5; that's a staggering 100 times brighter than Venus at its brightest and able to brighten the night sky as much as a fat crescent moon. However, before we really start to crave such an object in the night sky, we should bear in mind the dangers posed by one to our very existence. Ironically, without supernovae and their ability to coalesce, as forming stars, from hydrogen, and ultimately produce heavy elements when they explode, life would not exist at all. Those heavy elements that formed the earth and human beings were produced by supernovae in the first place. But how close does a supernova have to be to actually threaten our health or survival? Well, first lets just get things into perspective. If you are the worrying type, start worrying about your weight, your fitness, crossing the road, and your stress levels before even considering anything as remote as a nearby supernova explosion. Events like asteroid or comet impacts and deadly radiation arriving from space attract the attention of the media and Hollywood because the possible eradication of all life on Earth is a *big* event, to say the least! However, statistically you are far more likely to be killed in a road accident or struck by lightning. Nevertheless, a prudent and civilized society does need to consider these cataclysmic possibilities, especially as, with an impact scenario, we are technologically advanced enough to conceivably divert an asteroid or comet coming our way. With a supernova explosion, we get no warnings though; but just what are the risks of a catastrophic event?

Obviously, the chances of our own solar system being disrupted by the blast wave from a nearby supernova are tiny. Supernovae are violent events, but the distances between neighboring stars are huge, and the chances of one of our nearest stellar neighbors becoming a supernova in the near future are totally negligible. There are no potential supernova candidates nearby anyway. The real risks are posed by X-rays, gamma rays, and cosmic rays sweeping through our solar system from a supernova within a few hundred light-years and certainly within 100 light-years. As with all issues like this, where there is only theory and no practical experience, experts disagree on the dangers. In addition, every supernova is different, and Type Ia supernovae pose different risks from the nonbinary system supernovae.

X-Rays

X-rays, that is, radiation with a wavelength between 0.01 and 10 nm (by comparison, the wavelength of visible light is 400–700 nm) are a particularly deadly by-product of a supernova explosion. Figure 5.1 shows the X-ray emissions from the Crab Nebula environment almost 1,000 years *after* the supernova explosion. The Chandra X-ray satellite is an especially powerful tool for analyzing supernova remnants as they are frequently strong X-ray emitters. Computer simulations of the energies released in Type Ia supernova explosions by Shigeyama et al. (A&AS 97, 223 [1993]) derive a value of 10^{42} ergs/second for X-ray emissions at their peak. An erg/second is defined as 10^{-7} joules/second or 10^{-7} watts. So 10^{42} ergs/second equals 10^{35} watts or one hundred billion trillion trillion watts, or, put another way, one hundred thousand trillion trillion megawatts! This is a colossal amount of energy and deducing how far away you need to safely be from such an event is largely a subject for debate. What, exactly do we mean by "safe" anyway? The only realistic radiation hazards from space that the scientific community ever concern themselves with are energetic solar flares, which pose a serious health hazard for astronauts and a minor one to the rest of us on Earth. Even energetic solar flares generally emit less than 0.5 ergs/second per square centimeter at Earth, some 150 million km away. If we were 500 light-years (4.8×10^{20} cms) away from a Type Ia 10^{42} ergs/second X-ray emission, the figure of 10^{42} would be divided by a spherical surface area of 2.8×10^{42} square centimeters (i.e., reducing it down to the level of a large solar flare). One might reasonably conclude from this that a Type Ia

Figure 5.1. A composite image of the Crab Nebula showing the X-ray and optical images superimposed. X-rays dominate the region, and the central ring is approximately 1 light-year across. Image: NASA/CXC/HST/ASU/J. Hester et al.

supernova might start to generate worrying levels of background X-rays if it were within several hundred light years of Earth. A number of Earth-orbiting satellites have measured X-ray fluxes from non–Type Ia supernova explosions, which are significantly lower than the Type Ia Shigeyama model. For example, the Type Ic supernova 1998bw in galaxy ESO 184-G82 had an X-ray luminosity measured by the BeppoSAX satellite equivalent to 5×10^{40} ergs/second, when the 140 million light-year distance is taken into account. This flux level appeared to decline gradually over the 6-month monitoring period, until it had faded to the equivalent of 1.7×10^{40} ergs/second. The Type IIP supernova 1979C, discovered in the galaxy NGC 4321, better known as M 100, was the 22nd brightest extragalactic supernova discovery of all time, when it was found, at magnitude 12.0, on August 19 of that year. Measurements made by the ROSAT satellite in 1995 and by the XMM satellite in 2001 imply an X-ray luminosity of just under 10^{40} ergs/second at 16 and 22 years after the event. While a 100-fold less intense than a value of 10^{42} ergs/second, it does show that X-ray emission can continue for years and decades after a supernova explosion, presumably due to shock waves encountering interstellar material in the supernova's vicinity.

Gamma Rays

Gamma rays, that is, radiation with a wavelength less than 0.01 nm, are the other major deadly by-product of a supernova explosion. Recently, there has been much discussion about historical global extinction events caused by the massive gamma ray bursts that are detected in our Universe. These events were so easily detected by the 1960s Vela satellites (and sometimes they have optical counterparts, too) that it used to be thought that they must originate within our galaxy. However, they actually occur in very distant galaxies, hundreds of millions or even billions of light-years from Earth. Extraordinary though it may seem, these gamma ray burst (GRB) events dwarf supernova explosions. Indeed, the biggest events are estimated to produce 10^{53} ergs of energy during a couple of seconds! To put this in perspective, our sun would produce the same amount of energy if it was active over a period of several trillion years! Theories for what could cause such a GRB event include the collision of two neutron stars and also the formation of a massive supernova called a hypernova. This latter event might result from the collapse of a supermassive star (of about 40 solar masses or more) into a black hole, liberating perhaps 100 times or more the amount of energy in a standard supernova explosion. In 2002, a group of astronomers led by James Reeves, at Leicester University, U.K., used the European Space Agency's XMM-Newton spacecraft to study the GRB that was detected on December 11, 2001 (GRB 011211). They recorded evidence for silicon, sulfur, argon, magnesium, and calcium; in other words, the elements normally associated with a supernova explosion. The Leicester team cited this as strong evidence that massive GRB events were caused by hypernova explosions, as a neutron star merger would, theoretically, be dominated by heavier elements like iron. Another group, a joint team at the Harvard-Smithsonian Center for Astrophysics and The University of Notre Dame, came to similar conclusions after studying the optical afterglow from GRB 011121 in November 2001. That GRB was reckoned to be a mere (!) 6 billion light-years away in light-travel time: quite a close one by GRB standards. Kris Stanek of that team remarked, in a paper submitted to the *Astrophysical Journal*, that the GRB afterglow faded quickly over several hours but then brightened a couple of weeks later and faded again, just as would be expected if the burst was part of a gigantic supernova.

Obviously such events, rare though they are, would be the most lethal from the point of view of life-threatening radiation. The damage to the earth's ozone layer from such a GRB event could be catastrophic. Research carried out by the Department of Physics and Astronomy at the University of Kansas in 2003 suggests that a GRB event in our galaxy may have caused the Ordovician extinction 450 million years ago (i.e., 200 million years before the dinosaur era). The theory suggests that a GRB, maybe as close as 6,000 light-years from Earth, could have devastated Earth's ozone layer for up to 5 years, allowing lethal solar radiation to kill off the smaller life-forms, thereby devastating the food chain. Even a normal Type Ia supernova some 1,000 light-years away from us could swamp Earth with more than 10,000 solar flares–worth of gamma radiation, but geophysicists think that more than 10^5 ergs/cm^2 of gamma radiation is needed to seriously damage the ozone layer. These figures suggest that a normal Type Ia supernova would need to be well within 3,000 light-years to really have any detectable effect on the ozone layer, whereas a much more powerful GRB hypernova might easily cause serious damage from 10 times that distance.

This is not a precise science, but from the above discussions we can see the magnitude of the X-ray and gamma ray supernova threat to life on Earth. For a normal Type Ia supernova, there are increasing levels of risk to Earth's environment starting as soon as the supernova distance moves within a few thousand light-years but becoming very damaging within a few hundred light-years. For GRB hypernovae the threat is much more serious, with an object even within 10,000 light-years posing a serious threat to all life on the planet.

Interestingly, in 2001, Jesus Maiz-Apellaniz and his team at the Space Telescope Science Institute found a supernova remnant in the group of stars known as the Scorpius-Centaurus association. The deduced age of that supernova corresponds with a puzzling layer of heavy isotopes in deep Earth core samples and also to a significant marine extinction 2 million years ago. At the time, Scorpius-Centaurus was around 300 light-years from Earth. Of course, this could just be a coincidence, we just do not know.

Nearby Imminent Progenitors?

Nearby giant stars that are likely to become core-collapse supernovae are not hard to spot. Such massive stars, even at distances of tens of thousand light-years, will be easy binocular objects. Within a few thousand light-years, they become naked-eye stars. One supermassive star that is a potential nearby supernova is the Southern Hemisphere's Eta Carinae (see Figure 5.2). In the 1840s and 1850s, an outburst of Eta Carinae (now believed to have a companion star) made it one of the brightest stars in the sky and that was just an outburst. When the 100 solar mass monster collapses into a supernova, it should be quite a spectacle. The blue supergiant star that is Eta Carinae probably produces more light in 6 seconds than our sun produces in a whole year; staggering!

Perhaps the best nearby supernova candidates in our night skies are Betelgeuse and Antares. Betelgeuse is known to every sky watcher as the red supergiant star in the top left corner of Orion (see Figure 5.3). Despite being 430 light-years away,

Figure 5.2. This stunning Hubble Space Telescope image shows a billowing pair of dust and gas clouds emitted by the supermassive star Eta Carinae. This star, a hundred times the mass of our sun, experienced a massive outburst more than 150 years ago when it temporarily became one of the brightest stars in the Southern Hemisphere. At a distance of 7,500 light-years, it will become a brilliant supernova when it finally goes bang. Image: NASA/ University of Colorado (J. Morse).

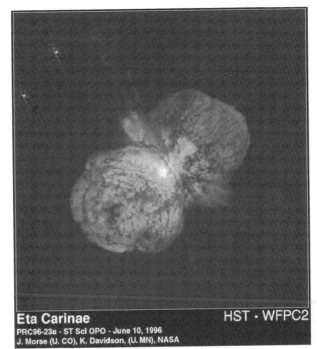

Eta Carinae   HST · WFPC2
PRC96-23a · ST ScI OPO · June 10, 1996
J. Morse (U. CO), K. Davidson, (U. MN), NASA

Figure 5.3. The constellation of Orion with the potential supernova, Betelgeuse, in the top left. Image: Jamie Cooper.

Figure 5.4. The star Betelgeuse clearly resolved as a disk in this ultraviolet Hubble image. Image: NASA/ESA/STSCI (Ronald Gilliland) and Harvard-Smithsonian CfA (Andrea Dupre).

it is visible to us as one of the brightest naked-eye stars, varying between about mag 0.1 and 0.9 and with an absolute brightness more than 10,000 times greater than that of our sun. If it was situated where our sun is, its gaseous outer atmosphere would extend as far as the orbit of Neptune. Indeed, the Hubble Space Telescope has actually resolved Betelgeuse as a disk (see Figure 5.4), little more than a tenth of an arc-second across in ultraviolet light. Betelgeuse will become a supernova in the cosmologically near future. The problem lies in knowing how far away in time this really is. We could see it happen tomorrow (in which case it would actually have occurred 430 years ago) or it might not happen for, say, 10,000 years. From the point of view of a star, being 10,000 years from the end point is practically a few weeks away in human life-expectancy terms. From our perspective, it is not something that is likely to be imminent. Unfortunately there is no way, just by looking at a star, to tell how far away the end point is, even when the star is a supergiant that will inevitably become a supernova at some point. All we can do is count up the most likely candidates within our vicinity. At 520 light-years distant, the red supergiant star Antares is another nearby, potentially imminent, supernova and arguably the second favorite to "go bang sometime soon." Fortunately, both these stars are hundreds of light-years away. If they were within a hundred light-years and shining almost as brightly as Venus, they would be too close for comfort.

With violent Type Ia explosions, the problem of spotting nearby hazards is much greater as the progenitor star is much fainter and harder to spot. A few years ago a Harvard University student, Karin Sandstrom, received a lot of publicity for noticing a star in our "galactic backyard" that might, according to somewhat hyped press reports, "wipe out life on Earth." The white dwarf star, called HR 8210, had actually been catalogued a decade earlier, in 1993, but Sandstrom noted it was a binary star that was probably close to its Chandrasekhar mass limit. At a distance of 150 light-years, it is a very close supernova candidate. However, its companion star does not appear to have yet expanded into a red giant and dumped its outer layers onto HR 8210. That event could take hundreds of millions of years. So, far from being an imminent threat to Earth, it really poses no threat at all. It is the nearby unknown white dwarf binary systems close to their Chandrasekhar limit

Figure 5.5. The recurrent nova RS Ophiuchi imaged by Giovanni Sostero and Ernesto Guido.

that are more worrying. Admittedly, any likely candidates that close will already be identifiable in all the professional star catalogues, but how close their comparison stars are to spilling their outer layers onto their white dwarves and how close those white dwarves are to their mass limit is unknown.

In passing, it is worth mentioning that amateur variable star observers are keen on observing similar types of systems that flare up on a regular timescale. Called Cataclysmic Variables, or CVs, they also feature binary stars in which hydrogen is gravitationally sucked from a companion star onto a white dwarf, or, a white dwarf's accretion disk. In this type of system, dramatic outbursts occur on a regular basis (typically on a timescale of months or years) as the accretion disk hot spot flares up. So CVs are, in many ways, a bit like watching a Type Ia supernova not quite going bang. This will prompt the inquisitive reader to ask whether any CVs can actually go the whole hog and turn supernova. The answer to this is yes, if material is accumulating onto a white dwarf that is close to the Chandrasekhar mass limit. The famous recurrent nova RS Ophiuchi, which flared from 12th to 4th magnitude in 1898, 1933, 1958, 1967, 1985, and 2006, appears to contain a white dwarf whose mass is slowly creeping up toward the Chandrasekhar limit. However, as it lies at an estimated distance of 3,000 light-years, it is not a threat to Earth. At the 2006 outburst (see Figure 5.5), it brightened to 4th magnitude on February 13. If it had been a Type Ia supernova, it would have reached magnitude −10.

Supernovae: A Threat to Life on Earth

Part II

Observing and Discovering Supernovae

Chapter 6

Supernovae as Visual Variable Stars

Making a visual brightness estimate (known as a magnitude estimate) of a supernova is, in theory, no different from estimating the brightness of any other variable star. However, with even the brightest extragalactic supernovae in living memory being only 10th or 11th magnitude (excluding 1987A), the faint nature of these stars makes estimating their brightness a major challenge in all but the largest amateur telescopes. There is an additional problem, too. Supernovae live in galaxies, and the background brightness of the galaxy can often complicate the magnitude estimate, unless the supernova's offset from the galactic center is considerable. For supernovae close to a galaxy's core, an accurate visual magnitude estimate can be almost impossible. However, despite these problems, dedicated visual observers, like the U.K.'s Gary Poyner, can make hundreds of magnitude estimates of such objects throughout the course of a year (see Figure 6.1).

Visual Magnitude Estimation

Many readers of this book may never have attempted a variable star magnitude estimate, so a complete grounding in the subject is worthwhile at this stage. To estimate the brightness of any variable star, whether a naked-eye, binocular, or telescopic object, it is necessary to compare it with other nearby stars of a known magnitude. Obviously, the comparison stars themselves must not vary in brightness, or all hope for a useful measurement is lost. Fortunately, for established variable stars, a reliable photometric sequence is almost always available, and organizations such as the BAA (British Astronomical Association), TA (*The Astronomer* magazine) and AAVSO (American Association of Variable Star Observers) can supply charts for hundreds of popular variables (see Figures 6.2 and 6.3). The relevant Web sites are listed in the appendix.

It can be tempting to simply use a planetarium software package to produce a homemade chart that you can use at the telescope. While this is perfect for making your own customized finder charts, it is very dangerous for making the magnitude estimate. Planetarium packages often have highly erroneous star magnitudes (the Hubble Guide Star Catalogue is full of them), and they do not tell you if a comparison star may be highly colored or slightly variable. A photometrically verified magnitude sequence should always be used where possible and the chart title/origin (e.g., AAVSO) and issue date/status should always be recorded. Photometric

Figure 6.1. The prolific variable star observer Gary Poyner who has made more than 200,000 magnitude estimates from his light-polluted sight in Birmingham, U.K. Image: Gary Poyner.

sequences are occasionally revised at a later date, so a complete record of which stars and which chart you used for your magnitude reduction is essential.

Charts for variable star observing come in a variety of formats, depending how large your telescope is and what magnification you are using. For Schmidt-Cassegrain users, they are invariably available in a left-right flipped format (i.e., to take account of the mirror diagonal often used with these instruments). Obviously, a variable star chart should be optimized to enable the field of view to be recognized and the star in question located. For my own observing, I used to find that having a homemade star chart that precisely matched the widest eyepiece field and showed stars to the same limiting magnitude as that first glimpse was essential. When you initially look through a telescope eyepiece, you are rarely perfectly dark-adapted. In addition, what you are trying to do is to identify star patterns. If you can only see the brightest stars on your chart, you may easily find a bright pattern of stars that looks surprisingly like a fainter pattern on the chart. When you go to the eyepiece, you must have total confidence that you know which way up is the north point and that the field of view exactly matches the chart field. For very faint stars and supernovae, you may well need a couple of charts (i.e., a wide-field custom chart to find the field and a narrow-field, high-magnification accurate chart for making the magnitude estimate). It is surprising how few faint stars you can see if you have only been dark-adapted for a matter of minutes. If you do not have a "GO TO" telescope, mechanical setting circles fitted to an equatorially mounted telescope can give you enormous confidence that you have the right field. Even a declination circle on its own can be a very valuable aid, especially for finding objects in twilight.

The eye (Figure 6.4) is a remarkable detector, but it does take a while to reach full sensitivity. For many years I used a massive 49-cm (19.3″) Newtonian at my former residence in Chelmsford, U.K. The telescope was right outside my living-room door, so I was in action just a few minutes after leaving the house. However, I was far from being dark adapted by that time. Somewhat surprisingly, I found that finder charts showing stars down to a mere magnitude 11 were the best ones to use to locate the right field. After half an hour and, at very higher powers, I could detect stars of magnitude 15, but for that initial location of the target a wide-field chart to magnitude 11 was ideal.

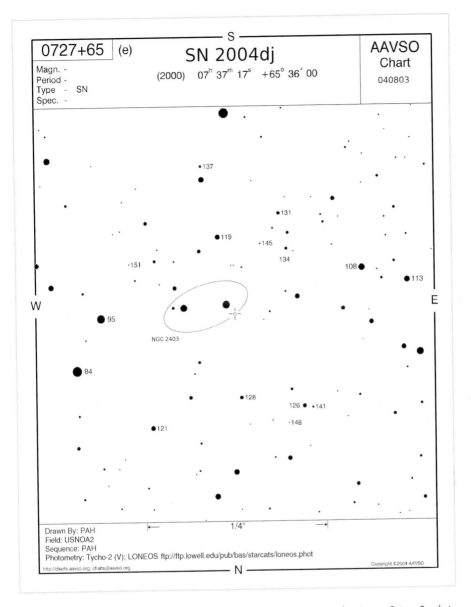

Figure 6.2. The AAVSO finder chart for supernova 2004dj, drawn by Aaron Price. South is at the top. Image: courtesy of Arne Henden/AAVSO.

Amateur astronomers have developed all sorts of tricks for seeing faint stars and eeking the very best performance out of their instruments. Obviously, the larger the telescope the better, and the darker your nighttime sky the better. But keeping your telescope in the peak of performance is vital, too. The most dedicated faint variable star observers have their telescope mirrors re-aluminized each year to maximize the light grasp. They are also careful in keeping their telescopes perfectly collimated so that, in good seeing conditions (when the atmosphere is stable) star images are crisp and not smeared by aberrations.

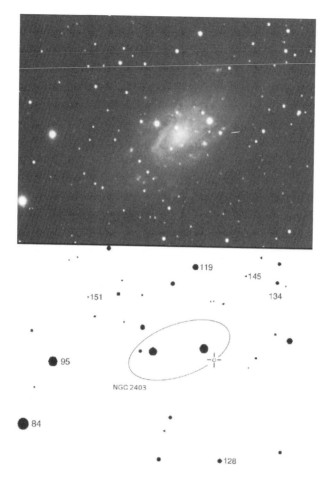

119

•145

•151

134

95

NGC 2403

84

•128

Figure 6.3. The AAVSO chart compared with an actual image of SN 2004dj by Jeremy Shears.

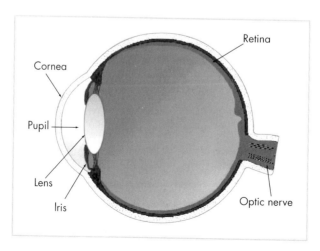

Retina

Cornea

Pupil

Lens

Iris

Optic nerve

Figure 6.4. A cross section through the human eye. It is still the most versatile detector of light and with only one disadvantage: it cannot take long exposures.

The Retina

There are plenty of visual tricks, too. Full dark adaption of the retina can take 40 minutes or more, so it can be prudent to start observing the brightest stars first in any observing session. Also, the retina itself is divided into two types of detectors; these are called cones and rods (Figure 6.5). The cones dominate in the central one degree of the field of view, called the fovea, where the highest full-color resolution is required. In a typical person with keen vision, a resolution of up to 1 arc-minute (1/60 of a degree) may be achieved by the cones in the fovea. While reading these very words your eye will be swivelling to place the words slap bang on the fovea's cones. But the fovea plays no role at all in seeing the faintest stars; for that we need rods, not cones. The electrochemical signals from the retina travel to nerves and on to the brain via cells known as ganglion cells. A ganglion cell may be mated to a single cone in the high-resolution fovea. However, it is a different story with the rods, where as many as a hundred may be served by one ganglion cell, to achieve the maximum signal-to-noise. As you move farther away from the center of the retina, the number of cones decreases and the number of low-light detecting rods increases. When the eye is plunged into darkness, the eye's pupil expands to its maximum size very quickly, and the chemical rhodopsin starts to increase in the retina. This chemical makes both rods and cones far more light-sensitive; by a factor of many thousands in fact. There is an optimum "most-sensitive" part of the

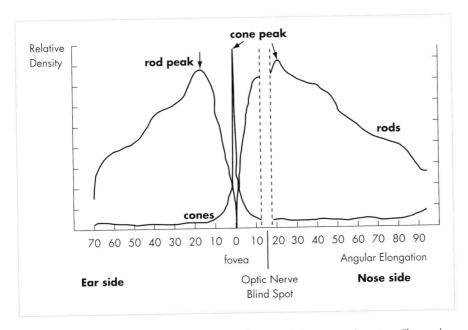

Figure 6.5. The number of rods and cones in a horizontal slice across the retina. The number of high-resolution, color-sensitive, bright-light cones peaks in the middle. However, the number of low-light, low-resolution, monochrome rods peaks some 20 degrees off center. The blind spot, where the optic nerve leaves the retina, interferes with the rod density on the side of the nose. However, the eye lens turns everything upside down and so the blind spot appears to be on the ear side of the eye. Therefore, faint objects are best viewed when they seem to be 12 degrees off-axis and toward the nose. Based on a diagram originally drawn by Osterberg in 1935.

retina where the observer should try to place the faintest stars. This point is about 12 degrees away from the fovea and, perhaps surprisingly, it is best if the 12-degree offset is arranged so the object seems nearer to your nose. This is because your eye has a blind spot on the opposite side, where the optic nerve leaves the retina. (Of course, if one were to slice the eye open, the optic nerve would actually be on the nasal side, as the eye's lens turn's everything upside down, but let's not confuse things!) Twelve degrees or so away from the fovea the eye is at its most sensitive. In fact, in darkness, it is 4 magnitudes (40 times) more sensitive at this point than in the field center.

For a beginner, mastering the averted vision technique can be very tricky. Instinctively, you want to place the object in the center of your vision, but you have to force yourself to look away from the target so the most sensitive part of the retina is used. You have to train yourself to ignore the central part of your eye when observing faint objects. This can prove almost impossible in a field where there are other bright background stars shining on your fovea. The main reason why averting your gaze by that 12 degrees (actually 8 to 16 degrees is fine) is especially tricky is that the highest resolution part of your vision is always in the center. You are just not used to studying anything in this manner in everyday life. While the naked eye might resolve an arc-minute with the color-sensitive high-resolution fovea, a much poorer resolution of 20 arc-minutes or so is typical at the monochrome edges of the visual field. Also, it might instinctively be expected that the "exposure time" of the eye was somewhere of the order a fifth of a second (i.e., not dissimilar to the human reaction time). While this is true for daylight observations, in darkness things are rather different. At the faintest levels it pays to stare at an object such that the flow of photons hitting the rods is sufficiently high enough for several seconds to trigger a definite "hit" in the brain. In practice, at a typical observing sight, with background light pollution, this level is usually reached when several hundred photons per second are hitting a group of rods. This staring action should not be interpreted as an exposure time as such. The eye is not a digital device like a CCD, but it certainly pays to stare at a faint object to see if it emerges. What the trained and dark-adapted eye sees, after studying an object patiently, is an impression not dissimilar to what a low-resolution CCD would capture with an exposure of a few seconds (see Figure 6.6). Of course, at these faint levels we can hardly ever make an accurate magnitude estimate of a star, all we can do is say we have seen it and maybe estimate its magnitude in relation to a slightly brighter comparison star. In elderly observers, the eye's pupil does not dilate as much as the 7 mm that the young eye is capable of. This means that a loss of light results at very low magnifications. The bundle of light rays from the eyepiece will have a diameter equal to the telescope aperture divided by the magnification. If this is larger than 7 mm, then light will be lost however young you are. Older observers will find that 5 mm is a more sensible figure to adopt. In other words, with a 300-mm aperture telescope, magnifications of 60 times or more are recommended. However, much higher magnifications are used for studying faint stars so the diameter of the pupil is really only a consideration at low powers.

Many beginners in this hobby seem to think that to see fainter you need a lower magnification. I guess this misconception comes from the fact that on bright objects like the moon, as you whack in a higher eyepiece the view just gets increasingly faint. However, different rules apply for the faintest objects. For a start, contrast becomes just as important as brightness. Only the luckiest amateur astronomers enjoy a really dark sky, and for many town dwellers the night sky has

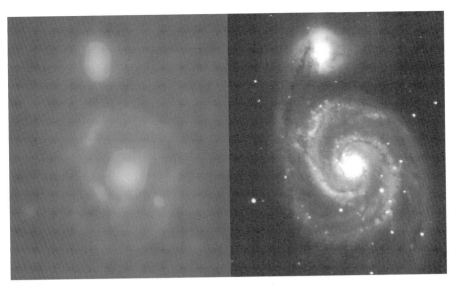

Figure 6.6. A CCD image of Messier 51, the Whirlpool Galaxy, by the author. On the left is a crude simulation (also by the author) of how the galaxy appears to the experienced visual observer using a large amateur telescope.

a horrible orange glow, from the thousands of nearby streetlights. As you increase the telescope's magnification, the background skyglow becomes much dimmer and point sources like faint stars (typically spanning a few arc-seconds in diameter due to atmospheric turbulence) start to cover maybe just a few more rods, convincing the brain that something is really there. Both of these factors enable the faintest stars to be seen more easily. Also, as the field of view becomes narrower, there is less chance of any really bright stars encroaching into the eyepiece field and dazzling the observer. At this stage, I would like to mention a point on which I am, admittedly, puzzled. Most knowledgeable amateur astronomers will tell you that a telescope's focal ratio has no bearing on how bright the background sky appears in a light-polluted area. The sky background will simply get dimmer and blacker as you whack up the magnification. If an f/6 and an f/4 Newtonian of the same aperture are used at the same magnification, the sky background brightness should look the same. Obviously, to match the magnifications, you would have to use two different eyepieces (e.g., a 4-mm eyepiece with the f/4 instrument and a 6-mm eyepiece with the f/6 instrument). However, if you ask really experienced visual observers who live in highly light-polluted areas what their opinion is, they will tell you that the slower f-ratio telescope (f/6 in this example) always makes the sky darker, even at the same magnification. To my mind this has to be an issue more related to scattered light in the tube than a straightforward optical issue. The f-ratio of a telescope should have no bearing on the background sky brightness for a specific magnification. The same number of photons will hit the same number of rods and cones when aperture and magnification are identical unless there is something different in the design of the two telescopes or the two eyepieces. Nevertheless, I have been assured this "slower is darker" rule applies by an observer who has made 200,000 magnitude estimates down to magnitude 17, so I have to take it seriously.

So, just how faint can the human eye reach? Well, for typical observers in good conditions the formula that is most often used is $2 + 5 \log_{10}D$, where D is the telescope diameter in millimeters. Thus, for 100, 200, 300, and 400 mm apertures, we should be able to see stars as faint as magnitude 12, 13.5, 14.4, and 15.0. In practice however, experienced observers can reach a full 2 magnitudes fainter than those values for stars at high altitude and on the clearest nights, even from urban locations. From black desert skies, even a 3 magnitude gain is possible for eagle-eyed observers! In fact, rods are so sensitive that they can actually detect single photons. In 1942, Selig Hecht proposed this because light flashes so dim that only 1% of rods were likely to absorb a photon were detectable by observers in experiments. Experiments by Schneeweis and Schnapf in 1995, using monkey rods, confirmed that single photons could trigger a response. The arrival of a few photons per second at the eye is the sort of rate that an observer using a 400-mm telescope might receive from a 20th magnitude star! Obviously, a pitch black sky is needed to test this theory. In a well-documented case, the amateur astronomer and author Stephen J. O'Meara spotted Halley's comet at magnitude 19.6 using a 600-mm reflector, at altitude on Mauna Kea, Hawaii, and while breathing bottled oxygen. Fortunately, the human retina has no electronic readout noise and does not need to be cooled!

Truly, the human eye is a quite remarkable detector. Even in the 21st century era of CCDs and Webcams, its versatility is extraordinary. It can cope with illumination levels from bright sunlight to starlight, spanning a hundred-million fold in intensity and even survey almost the whole night sky for meteors, following them instantly as they whiz across the sky. Amazing!

Magnitude Estimates and Terminology

In practice, experienced variable star observers use one of two methods to estimate the magnitude of a variable star. These are known as the fractional method and the Pogson step method. Although magnitudes are generally quoted to one decimal place (e.g., 13.1 or 13.7), in practice visual accuracies of ±0.1 magnitude (equivalent to ±10%) are usually achieved by luck, at least where really faint stars are concerned. If an observer thinks his magnitude estimate is likely to be accurate to within ±0.1 magnitudes, it is known as a Class 1 estimate. In other words, as good as it can be. The human eye/brain is not a photometric machine, it is really designed to cope with extremes of brightness and has a largely logarithmic response. Magnitude estimates are just that, estimates to, at best, the nearest 10% and usually far worse. This, though, is not a real problem, as many variable stars show considerable variations and even a measurement of ±20% accuracy is very useful. The precise definitions of the classes of variable star estimate, as defined by the British Astronomical Association (BAA) Variable Star Section (VSS), are as follows:

Class 1: Very confident of the estimate made under ideal conditions and confident of an accuracy of 0.1 magnitudes.

Class 2: Less confident than 1, maybe cloud or tiredness or stray light interfering. An accuracy of 0.2 magnitudes.

Class 3: Observation made under extremely poor conditions; variable just glimpsed a couple of times. An accuracy of 0.3 magnitudes or less.

Speaking personally, I think the only way I could achieve a Class 1 estimate would be if the two comparison stars were 0.1 magnitudes brighter and fainter than the bright variable star I was estimating. But let us now have a look at the actual estimation techniques.

Fractional Estimates

The fractional method is very easy to understand. Imagine you have a supernova of roughly magnitude 14. You have already acquired a star chart from the AAVSO or BAA and you spot two suitable comparison stars on that chart. One is labeled E and has a magnitude of 13.4. The other is labeled F and has a magnitude of 14.2. Armed with the chart and a dim red light, you approach the telescope drawtube, replacing the low-power eyepiece you used to find the field with a high-power one. After locating the supernova and the two comparison stars, you turn the red light off and stare at the field, using your newly acquired "averted vision" techniques. Imagine that the supernova appears to be a bit fainter than star E, but a lot brighter than star F. In fact, it is roughly a quarter of the way in brightness from star E to star F (which makes it about magnitude 13.6). Confident of this measurement, you turn your dim red lamp back on and make the following observation in your log book: E (1) v (3) F = 13.6.

It is vital to record the fractional estimate in full like this and *not* just the actual value. Why? Well, as we saw earlier, occasionally the comparison stars used to estimate magnitudes have their magnitudes revised. This can happen if a chart has been hastily issued after a supernova discovery. If one of the comparison star magnitudes is changed and you know which stars you used to make the estimate, then all is well and good. You can go back a few weeks later and revise the estimate if needed. However, if the whole fractional estimate was not recorded in the above manner, just the value, all hope is lost and the observation is, at best, dubious, at worst, null and void. Of course, if the supernova had been indistinguishable in magnitude from star E, you could simply record a direct comparison magnitude in the form: Supernova 2008xx = Star E = 13.4.

Pogson Step Estimates

So, we have looked at the fractional method; now for the Pogson step method. This sounds a bit more daunting, but I am somewhat mystified by what merits the attaching of a man's surname to it. The Pogson step method is, quite simply, just using a single comparison star and estimating how many tenths of a magnitude your variable or supernova is above or below it in magnitude. I would like to say more, but there is nothing more to say, except, perhaps, give an example. Let us, once more, take our magnitude 13.6 supernova. But this time we only have star E, of magnitude 13.4, as a comparison. So, we estimate the mag as E − 2, that is, 2 steps *fainter* than 13.4, or mag 13.6 (remember, magnitudes get bigger as stars get fainter). Most observers will only use this method if they are short of comparison stars or if they are very experienced at estimating tenths of a magnitude. Whether you use either the fractional or the Pogson method, they are both pretty obvious. It is only the notation method that needs memorizing. One final point: what do you do if you simply cannot see the star but can see many of the comparison stars? The notation for this is simply the "<" symbol. So, for example, <E can be recorded at the telescope, translating to <13.4 in the reduced observation. In other words, the star was fainter than the faintest comparison star you could see on that night. A negative observation like this can still be of value. With supernovae, at some point the star will fade beneath everyone's threshold.

So what data needs to be included in your logbook? Quite a lot. Essentially you need: star designation; date and time; Julian date (see below); raw estimate [e.g., E (1) v (3) F]; derived magnitude (e.g. 13.6); observation class; star chart sequence and version; telescope used and magnification.

If the data is being entered on a computer, with a spreadsheet like Microsoft Excel, you may wish to enter other information as well. For example, a variable star observer who was interested in a whole range of objects might want to set up a code for each different type of object. Alternatively, the right ascension and declination might be stored in another column to allow easy sorting of the objects into positions in the night sky.

Julian Date

I have mentioned the term *Julian date* in the section above, so I had better explain it. Years, months, and days can be a bit inconvenient for the analysis of the behavior of a long-term variable. This is not the case for supernovae, as they simply brighten explosively, stay quite bright for a while, and then fade beyond visibility. However, other variable stars do exhibit periodic or cyclic behavior. It is very useful therefore to simply have a system where the time of the observation can be expressed simply in days and fractions of a day. Admittedly, in this computerized era a PC can easily allow for the length of the month and take account of leap days, etc. But before the 1980s, this option did not exist and so the Julian date system was adopted. This system is a purely historical one, with little actual relevance to astronomy. To digress for a moment, the Julian calendar changed to the Gregorian calendar in Roman Catholic countries after October 4, 1582, and the next day became October 15. However, in England the change was not adopted until September 1752. The change was necessary because the Julian calendar had too many leap years and so 10 days had to be subtracted just to correct the accumulated error. The Julian calendar has a leap year every fourth year, while the Gregorian calendar has a leap year every fourth year except century years not exactly divisible by 400. Either way, variable star observers are only interested in the number of days between successive observations and the Julian date system allows observations prior to 1752 to be included smoothly, without any mysterious calendar hiccups. Julian dates (abbreviated JD) are a count of days and fractions of days since noon Universal Time on January 1, 4713 BC. Before you get out your calculator though, the JD number is given in many astronomical almanacs and Web sites. A good online converter can be found at http://aa.usno.navy.mil/data/docs/JulianDate.html. Please note that, somewhat bizarrely, the JD decimal of a day is zero at noon UT, which flies in the face of conventional wisdom somewhat. Zero hours UT is twelve hours in the Julian day. Put another way, if the UT fraction of a day is 0.5, the Julian day fraction is 0.0. Readers of this book may prefer a reference point nearer to the time of publication. On 2006 January 1 at 12 hours UT, or 2006 January 1.5, the corresponding JD number was 2,453,737.

Bits and Pieces

Most hardened amateurs know that observing is about more than just a telescope and a chart. The observing environment needs to be friendly, instinctive, and hassle-free. Between 1980 and 2002, I mainly used two extremely large and user-unfriendly telescopes with apertures of 36 and 49 cm. I was rarely in a comfortable position with either instrument and, more often than not, I was up a stepladder with a twisted neck. The worst situation was when I was observing an object near (but not quite at) the zenith, when the lower part of the telescope tube prevented me from getting the ladder close to the telescope at all. An unfriendly telescope can really sap your enthusiasm. If there is one thing I have learned in more than 30 years in this hobby, it is that at night, in the cold, dark, and damp, an ergonomic telescope and workstation is essential, especially for visual work. Subzero temperatures are a real enthusiasm sapper and, it goes without saying that appropriate warm clothing, or the proximity of a warm room, are essential. If a telescope is positioned within a few yards of a house door and can be swung into action in a matter of minutes, the hobby becomes far more tolerable, even in the depths of winter. Personally, I find that some thermally lined boots (sometimes called "moon boots"), fingerless mittens, and a good warm balaclava are the most essential requirements. The extremities of the human body are the most susceptible to cold temperatures, as Arctic and Antarctic explorers who have suffered frostbite will testify! But one can overinsulate the main torso, leading to the observing session being quite a sweaty one if the observer is dressed up like a Christmas turkey. I have always preferred to have a nearby warm room to retire to for 20 minutes than to have to spend 15 minutes dressing up and dressing down before an observing session.

For visual observing, a rugged observing logbook is essential, too (i.e., something that won't fall apart or blow away in a breeze and that will last for a year or so of steady observing). A variety of reliable pens, ballpoint pens, felt tip pens, and pencils should be available, too. In cold or damp weather, 90% of writing implements refuse to make any mark on the paper. If a variety of pens are available, they can't all freeze or get damp! A solid writing surface, at waist height, should also be available, as well as a chair. Continually bending down to write on a pad lying on a concrete floor does neither my back, nor my knees, any favors.

I have already mentioned a dim red lamp (because red light dazzles the eye far less, and the night vision remains intact) and a red bicycle rear lamp works fine. However, many observers prefer a head flashlight of the sort worn by miners (with a strip of red plastic over the lamp), and my personal preference is a device made by Black & Decker, called the Snake Light. This is a flashlight attached to a long, flexible (but not floppy) hose with a battery compartment at the end. The batteries act as a nice counterweight to the flashlight end. The Snake Light hose can be wrapped around the observer's neck with the battery on the chest and the flashlight itself positioned between shoulder and ear. This system works well and does not seem to bang into the eyepiece end of the telescope as a miner's lamp tends to do.

The observer needs somewhere to store his star charts, too, where they will not get damp. An ideal solution to this is a looseleaf ring binder folder with transparent page holding pockets.

In practice, how many supernovae can you expect to observe during the course of a year? Well, I would hope that anyone reading these words and getting bitten by the variable star bug would observe far more stars than just faint supernovae. Novae, dwarf novae, and cataclysmic variables are just as interesting, if less dramatic. However, every year there are about a dozen supernovae brighter than magnitude 15, so with a big amateur telescope, whatever hemisphere you live in, you should not be starved of supernovae. However, rather predictably, with a CCD camera you can monitor and image far more, and we will deal with CCD photometry of supernovae in the next chapter.

Chapter 7

Supernova Photometry and Light Curves

Despite the fact that many visual magnitude estimates of supernovae are made every year, the amateur astronomy CCD revolution has brought precision photometry within the reach of the backyard amateur. Unlike the human eye/brain combination and its struggle to achieve better than ±20% accuracy with photometric measurements, the CCD can reliably achieve ±10% even in a beginner's hands and ±5% with a bit more care. Indeed, in the hands of an expert, much higher precision is possible. Beyond all this, though, the CCD detector has one other huge advantage: it can integrate light for many minutes, leading to a far greater signal-to-noise ratio than is possible with the human eye. In addition, images can be stored for eternity allowing a full reevaluation if any doubts are cast on the original magnitude estimate. Despite these obvious advantages, though, it is amazing just how faint an experienced visual observer, like the U.K.'s Gary Poyner, can go. With a 35-cm Schmidt-Cassegrain, he can just detect stars as faint as magnitude 16.8.

There are a number of critical factors that need addressing before a CCD image can be used for an accurate photometric measurement. I will briefly discuss these issues before dealing with them in more detail later. The first factor is one of color. A CCD detector has a far wider spectral range than the human eye. It can see deep into the near infrared and into the ultraviolet, too. For astronomers, this is a problem as stars can behave differently in, for example, the infrared compared with their performance in the normal human visual range. Professional astronomers take measurements in specific wavebands called U, B, V, R, and I (ultraviolet, blue, visual, red, and infrared), which enables them to precisely define a star's performance at each color and to calibrate their photometry accurately (see Figure 7.1). The V band corresponds with the center of the human visual range (i.e., to the color green). At first, the use of filters may seem a backward step. A CCD is so much more sensitive than the eye or film, but then we have to make it less sensitive by slapping filters in front of it. However, it is all in a good cause: the quest for scientific accuracy.

The second critical factor relates to the linearity and usable range of a CCD. Much CCD photometry is based on comparison with a single star. Both the comparison star and the star being measured should be well above the noise floor of the image and well below the (potentially nonlinear) saturation point. In general, CCDs are very linear devices with the charge collected in each pixel being accurately related to the number of photons collected. However, in devices with antiblooming gates (ABG), appreciable nonlinearities creep in as the pixel "well"

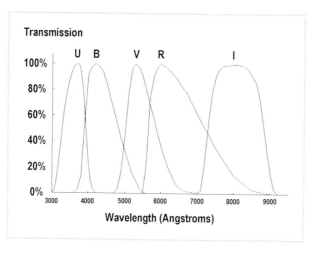

Figure 7.1. The standard photometric Bessell transmission curves used by professional astronomers. As described in the text, U, B, V, R, and I represent ultraviolet, blue, visual, red, and Infrared wavebands, respectively.

becomes half-full (i.e., halfway to saturation and white-out). Antiblooming gates are generally not found on purely scientific CCD detectors. Their sole role is to drain away excess charge, much as a storm drain copes with a surge of water. This draining activity prevents any likelihood of a very bright star "bleeding" down the image and so is vital for commercial applications where a pretty picture is the aim and not a scientific measurement. When an antiblooming gate first starts to work, it is siphoning vital data away from your measurement of a bright star in the field, *not* what we want. Rather predictably, non–antiblooming gate detectors are designated by the abbreviation NABG. In short, ABG CCDs should only be used where the comparison and target star are filling no more than 50% of the available range of the detector.

As a first step, the linearity of a CCD detector can be verified by experiments using, for example, three stars for the test (i.e., a faint comparison star, a midrange target star, and a bright comparison star). This was the method I used for my very first CCD camera at the start of the 1990s. A much quicker method of verifying your CCD camera's suitability for photometry is simply to join a like-minded group of people, such as the BAA Variable Star Section (http://www.britastro.org/vss/) or the AAVSO and join a user group where ideas and tips can be freely exchanged.

When using two comparison stars to measure an initial test target, they should, ideally, be just above and just below the target star's magnitude, as this will allow accurate calibration of the camera's linearity for a given exposure. If your camera software truly allows you to monitor the digital output from the A/D converter (prior to any potential software "fixes" to correct camera anomalies), you can be more confident that you are aware of your system's own deficiencies. It is also imperative that no image processing routines (except basic dark frame subtraction and flat-fielding) are carried out prior to the photometric analysis. The crude, but reasonably effective, technique I carried out to calibrate my original CCD Camera back in 1992 is detailed below. This was used in conjunction with a "V" (visual band) filter suited to the system, and the calibration procedure was intended merely to deliver an accuracy of ±0.1 mag. It was not intended to produce superaccurate photometry, merely to cope with linearizing results made with a definitely nonlinear ABG CCD.

Crudely Calibrating a Nonlinear CCD

The first step in calibrating a CCD or making a measurement with one is to make sure that all the stars used in the magnitude estimate are well below saturation and there is one photometric comparison star less than a magnitude brighter than the target and one less than a magnitude fainter. Saturated comparison stars are useless, and even those more than 50% saturated, prior to dark frame subtraction, should be avoided. As I've already stressed, as soon as a star approaches the well capacity of an ABG CCD, the antiblooming gates, used to drain excess charge away, will make the star look fainter than it really is (i.e., the photometric relationship becomes nonlinear). CCDs without antiblooming gates are more sensitive to light and more linear, however, as I have already mentioned, they allow bright stars in the field to "bleed" light down the image. The second step in reducing a CCD magnitude is to carry out the dark field subtraction and flat-fielding routines. Flat-fielding can be critical if the telescope field is badly vignetted and comparison stars lie in the vignetted region. Dark framing and flatfielding are described later in this chapter. The third step is to determine the linearity factor (as I call it) of the CCD; we will call this x in this example. I would just like to reiterate that this is a crude and optional "first step" for beginners who are making their first CCD magnitude estimates and think they have a CCD equipped with antiblooming gates. Later in this chapter, we will look at CCD photometry with some popular packages from a slightly different viewpoint. The reader may well wish to look at those sections first, especially if he or she owns one of the packages and is totally sure that the CCD camera in use is linear.

Even the crudest CCD camera packages allow the user to select a small box or "aperture" a few pixels square on the screen and to measure the integrated charge value within that box (i.e., a figure equivalent to the total amount of electrical charge collected by those pixels). Access to this data is essential in the example below. We will call the bright comparison star A and the faint comparison star B. The linearity factor x across the dynamic range can be deduced from the formula:

$$x = \frac{\text{Mag B} - \text{Mag A}}{\log\left(\text{A-noise}/\text{B-noise}\right)}$$

where Mag A is the known magnitude of star A; Mag B is the known magnitude of star B; A-noise is the integrated charge reading for star A minus the integrated charge reading for the noisy background sky; B-noise is the integrated charge reading for star B minus the integrated charge reading for the noisy background sky. Let us assume that star A has a magnitude of 12.0 and star B has a magnitude of 14.0. In this case, in a perfectly linear system, the value (A-noise/B-noise) will be two magnitudes or 6.3 times, so $x = 2/\log(6.3) = 2.5$.

(The discerning reader will note that this is similar to the 2.5 × ratio associated with the standard magnitude law (i.e., five magnitudes = 100 and the fifth root of 100 is 2.511886432). In fact the 2.5 here arises from the fact that the reciprocal of the log of the fifth root of 100 is exactly 2.50. Incidentally, this tells us that to calculate the magnitude difference of two stars, we merely need to use the formula: mag difference = 2.5 times $\log_{10}$ brightness ratio)

At the saturated end of the scale for my original 1992 CCD camera, with its antiblooming gates in full and horrendous action, the factor x becomes 5, not 2.5, that is, two stars that differ in brightness by 2 magnitudes appear to differ by only

1 magnitude; hardly the photometric accuracy that we desire and a good reminder to stay well below saturation when using ABG enabled CCDs.

Once we have the factor x determined, we can use it to scale our actual photometric measurement on a target star. For a target star between A and B in brightness (or even one slightly brighter than A), one can use the following formula:

$$Mag = Mag\ A + x\log(A\text{-noise}/Target\text{-noise})$$

Alternatively, if you have a target star slightly fainter than B, try the formula:

$$Mag = Mag\ B + x\log(B\text{-noise}/Target\text{-noise})$$

I repeat, the procedures above are necessary to guarantee ±0.1 magnitude accuracy with a nonlinear or ABG-enabled CCD system. Photometry of much greater precision is possible with meticulous care, NABG CCDs, and with the sort of software packages in wide usage (i.e., AIP, CCDSoft and Maxim DL).

Photometry in Detail

We have already seen that color and linearity are major factors that affect the accuracy of a CCD magnitude estimate. These are not the only issues, though. Measuring the brightness of a star with a CCD full of pixels is not dissimilar to measuring the amount of rainwater in a field full of buckets. If some of the buckets (pixels) are literally overflowing (with a bright star's charge) then we have an inaccurate measurement. If there are only a few drops of rain in the bottom of the bucket, the measurement will be almost impossible to make accurately (equivalent to measuring a faint star) especially if the bucket was damp to start with. Also, if some of the buckets are covered with obstructions, we have even more problems. In the separate panel labeled "Bits and Brightness," I have digressed a bit into the binary numbering system and how it applies to CCDs. But if you are happy with this concept, feel free to skip this panel.

Bits and Brightness

CCDs measure the amount of electrical charge in their pixels in quantities called bits, simply because computers and digital electronic devices only understand ones and zeroes. Perhaps surprisingly, to a non-mathematician, any number can be represented by ones and zeroes if each digit represents a number twice as large as its predecessor. Digital "bits" are part of the binary (base 2) numbering system, whereas humans, having 10 digits on our hands, work to a base of 10, not 2. In the binary system the decimal numbers 1, 2, 4, 8, 16, 32, 64, and 128 are represented by the binary numbers 1, 10, 100, 1000, 10000, 100000, 1000000, and 10000000. These numbers, and larger powers of 2, can be used to represent any number you like. A standard 16-bit binary number is simply one that has 16 digits and it can represent any number between zero and 65,535, and therefore any brightness level from zero to 65,535 units. A few examples might illustrate the principle. Let us pick the number 65,000, at the top end of the scale. The full 16-bit binary scale from 0000000000000001 to 1000000000000000 represents the numbers 2^0 to 2^{15} or 1 to 32,768. So how do we represent 65,000? Well, 65,000 is the sum of the following 2^n numbers: 32,768 + 16,384 + 8,192 + 4,096 + 2,048 + 1,024 + 256 + 128 + 64 + 32 + 8. The 2^n numbers 512, 16, 4, 2 and 1 are not required to make the number 65,000.

Bits and Brightness (Continued)

So, in binary, 65,000 is represented by the number 1111110111101000. If all 16 digits are "on," the number 1111111111111111 represents 65,535 (i.e., the largest number in a 16-bit system). In a 16-bit CCD camera, the stars just below the white-out saturation point, after the charge has been measured and converted from an analogue quantity to a digital number, will be just below this 65,535 level. You may see a measurement quoted as 60,000 ADUs, which simply stands for analogue-to-digital units; in other words the amount in the bucket has been measured and converted to a digital number. The faintest stars measurable, in an exposure of a few minutes, will typically, when initially measured, have brightness values of several thousand, or maybe even 10,000 ADUs, simply because they are sitting on top of a wall of charge composed of light pollution and electrical thermal noise. When the background brightness they are sitting on top of has been accounted for, the faintest stars themselves may have ADU values below a thousand, but with considerable error, unless the background brightness and any vignetting have been perfectly accounted for. Of course, a star in practice does not occupy a single pixel, but this aspect is dealt with in the main text.

Before an image of a variable star is exposed through the appropriate filter, a few preparations need to be made. First, the star you are measuring needs to ideally fall in the midrange of the image. In other words, for a 16-bit measurement, with a maximum range of 65,535 ADUs (analogue-to-digital units), on an NABG camera, if the target star and comparison star are, say, giving readings of between 20,000 and 40,000 ADUs, that is excellent. In practice, things will never be quite that easy and, especially where a supernova has faded below the faintest star on the photometric sequence, it is quite possible that the faintest comparison star in a sequence may end up registering 50,000 or more ADUs with the supernova down at the 5,000 ADU level in a deep exposure. In other words, the supernova might be 2.5 magnitudes fainter than the comparison star, so the accuracy will suffer. Some prior knowledge of the number of ADUs produced by a given exposure through a specific filter on a specific star is needed, and this can only come with experience. It is of little use taking an image of a supernova if you find the next day that everything is overexposed. However, I have some experience, so here are a few examples. Let us take a typical 120-second exposure with my Celestron 14 working at f/7.7 with an SBIG ST9XE CCD camera employing a KAF-0261E CCD. This is an NABG detector with a linear response and the telescope lives at a dark country site. The image scale, with the ST9XE's 20 micron pixels, is 1.5 arc-seconds per pixel. In a 120-second *unfiltered* exposure, I generally find that stars of magnitude 12 or brighter are just starting to saturate a few pixels at the center of each of those bright stars, that is, they are brighter than 50,000 or 60,000 ADUs (exact saturation occurring at 65,535 ADUs). Trying to measure a star of magnitude 12.5 accurately is certainly safe and possible; anything much brighter than that is getting too bright and dangerously near to saturation. Stars of magnitude 13.5 or so lie roughly in the midregion of the unfiltered CCD's range in a 120-second exposure. In the same exposure time, from my dark site, using the equipment above, on a crystal clear night, I generally get a noise floor of about 1,200 units composed of a sky background component and the thermal noise component from the camera. In the nature of noise, it is noisy, that is, it is not just a subtractable smooth solid level; there is a random noisy mess of dark gray pixels with a variation of a dozen or more ADUs superimposed on the wall of subtractable sky and thermal signal. It goes without saying that as faint stars get near to this random noise ripple, it is impossible to measure their magnitude with accuracy. Again, referring to my own images, pixels under stars of mag 15 or 16 are still detecting a healthy several thousand ADUs of star plus noise, but fainter than magnitude 17 the stars are contributing less additional light than the sky background they are sitting on top of. By the time we reach 19th mag, recorded stars are just an extra 10% addition on top of the noise wall and the random noise is badly eating into our confidence of the measurement. They can still be measured, yes, but with a poor accuracy, for example, mag 19.3 ± 0.3, compared with, say, a star of mag 14, which might be measured fairly easily as 14.1 ± 0.05, with some care and the same equipment. Essentially, with my unfiltered system, at 1.5 arc-seconds per pixel, the noise ripple can be thought of as being similar to that of an underlying carpet of random 20th or 21st magnitude stars, with everything else sitting on the top.

So, what have we learned from the last few paragraphs? Well, with a 355-mm aperture telescope and a sensitive CCD camera, working at 1.5"/pixel, unfiltered photometry of supernovae between, say magnitude 12 and 18.5 can be carried out with a 120-second exposure and some care. However, it is filtered work that is of most use to professional astronomers and, for this, we really need to knock at least 1.5 magnitudes of sensitivity from the above calculations. As a rule of thumb, accurate, filtered photometry on supernovae between mag 10.5 and 17 could be undertaken with the same equipment and exposure time. Obviously if we want to go fainter, a longer exposure time will work in our favor as although the sky background and thermal noise will then increase, the amount by which the faint stars rise above the random noise will increase, too (i.e., we will have more ADUs to safely play with). It might seem incredible that an amateur looking visually through the same instrument might be able to make any estimate of a 15th or 16th magnitude star at all. Why does the CCD need a 120-second exposure but the amateur just looks for a few seconds? The answer is that at these faint levels the eye can only just glimpse the object and hazard an educated guess. In many cases of stars "on the limit" variable star observers have asked me to image a field to confirm what they have seen because they were unsure if they really glimpsed the target. In quite a few cases I have found nothing there; sometimes the eye can play tricks. Also, remember that we are pinning down a precise photometric measurement with a CCD, typically aiming at a 5% precision. In contrast, a top visual observer is trying just to glimpse an object at his or her absolute limit within a confusing sea of light pollution.

FWHM, Apertures, Darks, and Flats

Hopefully, by now, I have conveyed some sense to the reader of the sources of error that arise when trying to measure the brightness of a star; color, detector linearity, and accumulating a strong signal without saturating the detector, are all crucial. However, we still have not looked at the nitty gritty of the math behind the measurement. Stars are *not* point sources, and their light spreads out over many pixels on a CCD detector attached to a modest astronomical telescope; in fact the eye, studying a CCD image often misses the fainter outer halo, just seeing the bright core. For a really accurate photometric measurement, we need to collect all of the star's light for measurement, but we do *not* want to collect light from nearby stars that might be very close to the star being measured. Often one encounters a situation where there are some very faint, almost invisible stars close to the star being measured. If their light is collected, the resulting magnitude estimate could be inaccurate.

A couple of technical terms need explaining at this point. These terms are *aperture* and *FWHM*. Now you may think you know all about aperture. It is the diameter of the telescope's mirror or lens, right? In the photometric context, no, wrong. In stellar photometry, the aperture is the diameter of the ring within which the starlight is being collected. A further term that is sometimes heard is *annulus*, that is, a measurement ring surrounding the star in question, solely composed of background sky brightness and thermal noise. This is used to calculate the height of the noise wall on which the star is sitting. So, as well as being very careful about deciding just how big our aperture is, to collect, say, 99% of the star's light, we have to have an accurate way of measuring the background brightness in magnitudes/square arc-second. Again, we do not want faint stars to confuse the background measurement either. Before this starts to sound too daunting, *don't panic*, most software packages do all this for you, but they still require a bit of intelligent user input. The sky background does not have to be measured by sampling the area just outside the aperture though. If this is polluted by other stars, another nearby patch of barren sky can be used for a "sky aperture" measurement. Typically, the annulus radius is about twice the aperture radius. The bigger the annulus the greater the statistical accuracy of the sky background measurement, *but* there is also a greater risk of background stars polluting the field, so nothing is fixed. You sometimes need to intelligently adapt your techniques to the star field in question.

Imagine a graph of a star's brightness as seen by the CCD pixels, showing the intensity along the y axis and the distance in fractions of an arc-second from the star center along the x axis. With perfect seeing and tracking, optical diffraction effects will produce a stellar profile like a smoothly curved mountain surrounded by a series of ditches, with the peak intensity at the top and the lower intensities at the base. Secondary maxima, each one a bit smaller than the last, will spread out from the star like ripples on a pond after a stone has been dropped in. Even allowing for poor seeing and tracking errors the profile will still be like a mountain gradually merging into the surrounding terrain. Determining where the faintest ripples or slopes merge into the still waters of the nighttime pond is very similar to deciding what aperture to use. The Gaussian-like curve of a star's profile can be defined in various ways that help the photometrist. The abbreviation FWHM is often seen in this context and stands for full width half maximum. This is the width of the star's profile at half its brightness maximum. In turbulent air, the effects of optical diffraction will be swamped by the spreading of the light due to atmospheric seeing, especially in large amateur telescopes. However, the profile will still have a roughly Gaussian shape. As a rough rule of thumb, many amateurs use a photometric "aperture" of roughly four times FWHM. In other words, take the star's apparent diameter at half the maximum brightness and multiply it by four. In practice, with a telescope like mine, at a low-altitude site, even the faintest stars leave obvious disks several arc-seconds across in long exposures and a sensible "aperture" to set is one of around 20 arc-seconds. With my 1.5 arc-seconds/pixel scale, this translates into just over 13 pixels diameter or an area of about 140 square pixels. However, the best plan is to acquire a star chart of a photometric test region and, by experimentation, deduce what is the best aperture for your system by measuring stars of precisely known magnitudes. In cluttered Milky Way fields, like the field of the prolific supernova producing galaxy NGC 6946, you may wish to use a slightly smaller photometric aperture. Of course, for the most accurate work, photometric sequences for the filter band in use should be used. In many cases, amateurs take images that are unfiltered and that roughly correspond to an "R" (red) filtered photometric sequence. Unfiltered measurements with the same CCD camera are still valuable, but to be scientific they should be labeled "unfiltered" so that any analysis takes this into account.

Images used for photometric purposes should never be subjected to the kind of advanced image processing techniques that amateur astronomers use to produce spectacular galaxy and nebula images. The *only* processes that should ever be used when photometry is being considered are dark frame subtraction and flat-fielding. Both of these processes make an image far more appropriate for photometry, whereas operations that alter image contrast, gamma, and the sizes of star disks (such as unsharp mask routines) totally wreck the suitability of the image for photometry.

Dark Frames

Many amateurs will be familiar with the process of taking dark frames. A dark frame is simply an exposure of the same duration, and at the same temperature, as the image of the star field, but with the telescope or CCD camera capped. Such an exposure records the non-sky component of the image (i.e., the thermal electronic noise and the bias level within each pixel). With many astronomical CCD cameras, the camera has an inbuilt shutter that blocks the incoming light during dark frame exposures. For cameras without a shutter, a cap can be placed across the telescope. Once a dark frame subtraction has been executed, the remaining image just shows the contribution from the night sky. The thermal noise, which increases with exposure time, and the electronic bias level have been eliminated, so starlight and, unfortunately, background skyglow, remain. For total flexibility and a complete calibration of the camera noise, scaleable to any exposure length, advanced amateurs sometimes record a "bias frame," too, that is, an exposure recording the bias levels that remain in an exposure of zero (or a very short) duration. Many amateurs also take multiple dark frames and average half a dozen or a dozen to get a more representative average dark frame. However, in practice, on many nights where you only have a matter of minutes between clouds, simply taking an image and an automatic dark frame of the same duration (and at exactly the same temperature) is the quickest practical solution. One might possibly think that you could get away with not taking a dark frame at all, as the whole image floor is simply lifted up by a certain amount by the electronic noise. However, in practice the variable star/supernova being measured and the comparison star are often in completely different corners of the CCD chip, and different pixels and CCD areas suffer in varying amounts from pixel-to-pixel variations and bias noise from the onboard amplifier. After subtraction of a dark frame, exposed at the same temperature as the main image, an image looks noticeably "cleaner" to the eye, especially in CCD cameras where there are significant pixel-to-pixel variations (see Figures 7.2 and 7.3).

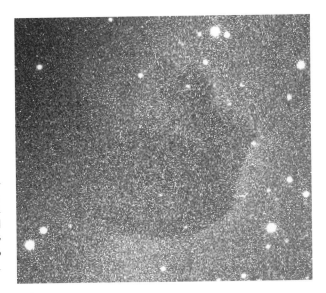

Figure 7.2. The Horsehead Nebula in Orion taken with a Celestron 14 at f/7.7, a 180-second exposure, and an SBIG ST9XE CCD, but with no dark frame subtracted. Image by the author.

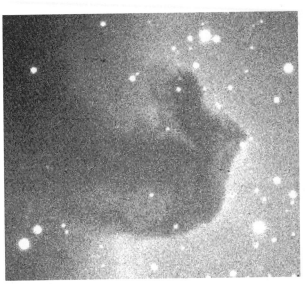

Figure 7.3. The same image as in Figure 7.2, but with a 180-second dark frame subtracted. Note the dramatic improvement.

Flat Fields

A flat-field, the other essential photometric calibration frame, has nothing to do with the camera electronics, but everything to do with deficiencies in the optical system (i.e., the telescope and any intervening lenses). Accurate photometry assumes that the whole of the CCD detector chip is receiving the light from the whole of the telescope mirror or lens. However, in practice, considerable variations can occur across the field being imaged. If the comparison star is in a part of the image receiving twice as much light as the supernova, the supernova may appear half as bright as it really is. Okay, differences this extreme might be rare, but *not* impossible. How do such variations in optical transmission occur? Well, between the main mirror/lens and the detector there are invariably various pipes through which the light passes through. These consist of secondary mirrors, narrow draw-tubes, and filter holders. In my own system, the light cone from my Celestron 14's primary is a fast f/2; this cone hits the secondary mirror and then, at a much slower f/11, heads toward a series of obstacle courses. First of all, it passes through the baffle tube and the Schmidt-Cassegrain drawtube. Then it hits a focal reducer that converts the f/11 beam to f/7.7. Then it hits a relatively narrow adapter prior to the filter wheel. It then passes through a filter housing only 28 mm in diameter and then through the narrow hole that leads to the CCD detector itself. Along the way, the cone of light is slightly vignetted at the edges and it also encounters numerous dust specks on the telecompressor, filter wheel housing, and CCD camera window. The end result is that, when highly contrast stretched, it is very apparent that the image center is far brighter than the edges and there are numerous "doughnut-shaped" ghosts visible. These doughnuts are, in fact, out of focus dust-specks.

The process of flat-fielding is often described as "dividing out" these illumination anomalies, a phrase that sometimes confuses the beginner who imagines that things like dust specks need "subtracting" not dividing. However, when you think about it, it is a ratio we are looking at here. If only half of the light is reaching the edge of the CCD detector, we need to divide those parts of the image by 50% to restore their full illumination (i.e., we need to divide our astronomical image by the flat field transmission). Modern software packages carry out the division process such that nothing in the resulting image is saturated when divided by a smaller number than 1.

So, how does one secure a flat-field image that accurately captures the vignetting and the dust specks but not the stars? Well, flat-fields are generally carried out with a relatively bright light source, and exposure times are very short. One good source of even illumination is a clear twilight sky, maybe 20 or 30 minutes after sunset. An exposure of a tenth of a second in such conditions will typically half saturate the CCD and give an excellent flat field (see Figures 7.4, 7.5, and 7.6). As we have

Figure 7.4. A twilight flat field obtained with the author's Celestron 14 at f/7.7. The dark corners are caused by vignetting of the light cone by the telecompressor, filter wheel, and adaptors. The small ring "doughnuts" are caused by out-of-focus dust specks.

Figure 7.5. A raw image, by the author, dark frame subtracted, but with no flat field applied.

Figure 7.6. The same image as in Figure 7.5 but with a flat field applied.

already seen with nonlinear ABG cameras, it is wise not to expose the CCD beyond 50% saturation. It is essential to capture the twilight sky flat-field as soon as the CCD camera's shortest exposure time can be used without the image saturating. If you wait 10 or 15 minutes longer, and it is darker, you will be using exposures of seconds and, even then, stars or star trails will start to appear, even with the drive turned off. Ideally, a set of flat field images, maybe a dozen or so, should be taken and averaged. It is important to realize here how easily a perfect flat-field can become an imperfect one. Specks of dust move about, even if blemishes on the lenses don't, and taking a CCD camera off a telescope and replacing it in a slightly different position certainly changes the dust speck situation and, quite often, the vignetting too. This is where a twilight flat-field obtained just before or just after the imaging session can be so valuable. It is a flat field of the same system you used to image the sky a few hours earlier or later.

Remember too, a dark frame should be taken for each of the flat-field frames too, so the essential image calibration routine will consist of the actual image of the supernova or variable star, with a dark frame of the same temperature and duration subtracted, followed by a division by the master flat field. This master flat field will, itself, consist of, maybe, a dozen flat fields minus their dark frames, all averaged.

With your system fully calibrated by the dark and flat fields, and using an appropriate filter, you should have no problem taking routine CCD magnitude measurements to ±0.05 magnitude accuracy when your target and reference star are within a magnitude of each other and not highly colored.

Photometric Filters

Professional astronomers measure magnitudes in accordance with internationally agreed standards (i.e., the sensitivity of their equipment is accurately defined at specific wavelengths). This is absolutely essential as, for precision work, you need to know that your CCD detector/filter combination is measuring exactly the same waveband as every other advanced amateur or professional. It is all a matter of calibrating your system across a wide range of colors. In the 1950s, Harold Johnson of the Yerkes and Macdonald Observatories created filtered measurement bands for the visual (V), blue (B), and ultraviolet (U) regions to suit the rather primitive blue-sensitive photomultiplier detector he was using. Later, he added red (R) and infrared (I) bands as he then had an enhanced red-sensitivity photomultiplier. Twenty years later, Cousins and Menzies (South African Astronomical Observatory) recreated the Johnson measurement bands using different filters on more sensitive detectors. Then, in the 1980s, Kron and Cousins modified the R and I system to better match the much more red-sensitive CCD detectors. In 1990, Bessell (Mt. Stromlo and Siding Spring Observatories) defined new filter bands that would recreate the entire UBVRI system for CCD detectors. Essentially, filter manufacturers, working with CCD manufacturers, now produce scientific-grade Johnson–Kron–Cousins UBVRI filter sets using the prescription defined by Bessell in 1990.

Predictably, the V-band is the band of interest for most amateurs. It approximates the visual band of the human eye so that CCD magnitudes through a V-filter are directly comparable with visual magnitude estimates. But the B and R bands are also of interest. Some variable stars vary considerably in the blue end of the spectrum, even more than they do in the visual band. CCDs are at their most sensitive in the R band, so this region is also of interest. The I band is of interest to specialist professional astronomers, but the U band is close to the limit of most CCDs spectral range. When the magnitude of a star is determined in both V and B passbands the B-V (B minus V) color index results. This color index can tell us a lot about the star; it can also be used to calibrate a photometric CCD system.

When choosing filters for a CCD camera, it is important to understand that the Johnson–Kron–Cousins bandpass boundaries define an ideal system and it is not possible to perfectly match a given set of filters to a CCD. The CCD will have its own response at specific wavelengths, and to derive the response of the system it is necessary to *convolve* the spectral response of the chosen filters with the spectral response of the CCD chip (see Figure 7.7). By convolve, we mean multiplying each point on the curve of the filters' spectral response, for every wavelength, with each point on the curve of the CCD's spectral response. But don't panic, because most CCD manufacturers will sell you, or recommend third-party vendors for, appropriate Bessell prescription filter sets to match their CCD cameras. As an example, SBIG can sell you specific photometric filters for their CFW-8 filter wheel or a complete photometric filter wheel matched to their cameras sensitivity (see Figures 7.8 and 7.9). Even if the filter-CCD match is not perfect, calibration correction factors can be applied to make it near-perfect if perfection is required. These correction factors can be verified by images of known test sequences in the sky for each passband. However, in practice, simply because supernovae are quite faint, most amateurs tend to carry out unfiltered photometry of all but the

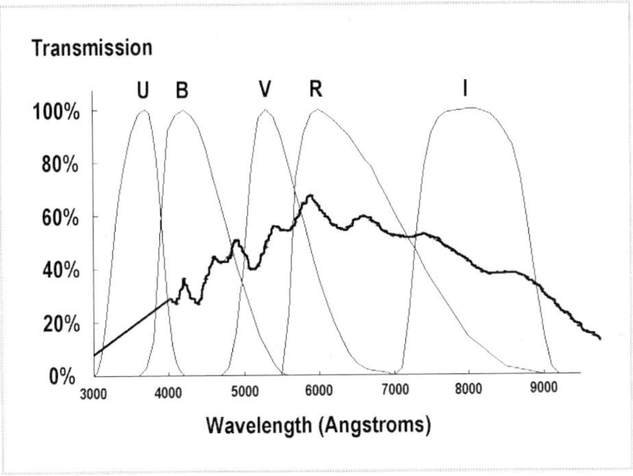

Figure 7.7. The transmission characteristics of the SBIG ST9XE's KAF0261E CCD chip overlaid on the UBVRI Bessell profiles. Note how the CCD sensitivity peaks in the red part of the spectrum.

Figure 7.8. The inside of the SBIG CFW8 filter wheel.

Figure 7.9. The author's filter wheel attached to the ST9XE camera.

brightest supernovae. Although this does compromise the absolute accuracy, it can, nevertheless, discern whether a supernova is Type II-L or Type II-P, for example, when a light curve is produced over many weeks. A useful paper to read in connection with photometry is the one by Arne Henden of the AAVSO at http://www.aavso.org/publications/ejaavso/v29n1/35.pdf

It is worth remembering that a V-filter will easily knock 1.5 magnitudes or more off your CCD camera's magnitude limit. When you add the fact that you need a decent signal from the star (but not enough to take it beyond 50% saturation), you need a surprisingly long exposure to get down to those 16th mag V-filtered stars that are just beyond most visual observers! Yet more proof of the formidable abilities of the human eye and brain, which can "rough guesstimate" a very faint star's magnitude in an instant.

Photometric Software

Various software packages exist for reducing stellar magnitudes from CCD images. Some amateurs use the software provided by the camera manufacturers, some use their own homemade software, and some use freeware programs they have acquired from the Web or from friends in their local or national societies. A few amateur astronomers, especially those with Linux-based operating systems (as opposed to Microsoft Windows), use the professional package IRAF. However, increasingly there seem to be three main packages that are routinely used by amateur supernova imagers, namely: Software Bisque's *CCDSoft*, Richard Berry/James Burnell's *AIP4Win*, and Herbert Raab's astrometric software

Astrometrica. The first time you use software to reduce some magnitudes from a test photometric field, you may well be disappointed by the apparent errors in your measurements, even after you have subtracted good dark frames and divided the image by a flat field. Remember this is a precise science, but there are many sources of potential error. First, if your image is not taken in the same filter band as the comparison sequence, then anything can happen. A star that is relatively bright in the near infrared will look appreciably dimmer in a V-band image than in an unfiltered amateur CCD image. Also, the choice of photometric aperture and sky background sampling annulus can be critical to the measurement, especially if faint stars lurk in either region. Typically, the annulus radius is about twice the aperture radius. I will reiterate what I stated earlier: the bigger the annulus the greater the statistical accuracy of the sky background measurement, *but* there is also a greater risk of background stars polluting the field. In addition, do not assume that even a professionally derived photometric sequence is guaranteed to be accurate. On a regular basis, astronomers discover that stars in photometric sequences are variable on a small scale or on a long time period. The whole science of photometry is full of pitfalls. For precise work you need to choose a reference comparison star that is proved to be photometrically precise and in an area of sky where there are no fainter stars close by. The comparison star should also be as close as possible to the variable star's magnitude. As soon as there is a difference of one or two magnitudes between the reference comparison star and the star being measured, the errors really start to creep up. In addition, and especially if you have a linear NABG detector, a nice bright image of the target object (but below saturation) will vastly increase the precision with which a measurement can be made, compared with a short exposure where reference and target star are low down in the noise. So do not be disappointed by your early results and do not feel embarrassed by submitting an unfiltered measurement. In any field of science, as long as you have recorded everything relevant to the measurement, you have contributed a scientific observation. When doing unfiltered work, extra checks on the accuracy are advisable. The easiest way to achieve this is to use several comparison stars to obtain the estimate and to see how the different comparison stars compare with each other and what results they give for the target star.

CCDSoft

The procedure for making approximate photometric measurements in Software Bisque's *CCDSoft* is relatively painless if you have a good photometric star chart for the object to hand. Simply click on the "photometry set up" icon and enter your telescope aperture, f-ratio, and pixel size. You then have to select whether the seeing was excellent, good, fair, or poor (see Figure 7.10). You then click the photometry "reference magnitude" icon and when you click on a star of known magnitude, simply enter the magnitude in the box. Then the third photometry icon (labeled "determine magnitude") can be selected. Clicking on your target star then gives you the magnitude of that star relative to the reference magnitude. If you have Software Bisque's planetarium package *The Sky* installed on your PC, other options are available to you, as both packages working in harmony can identify your CCD image star field and call up the magnitude data on all the stars in the field. However, some caution is necessary here because you will want to know that the stars you are using are not variable and have been catalogued accurately. In the case of the

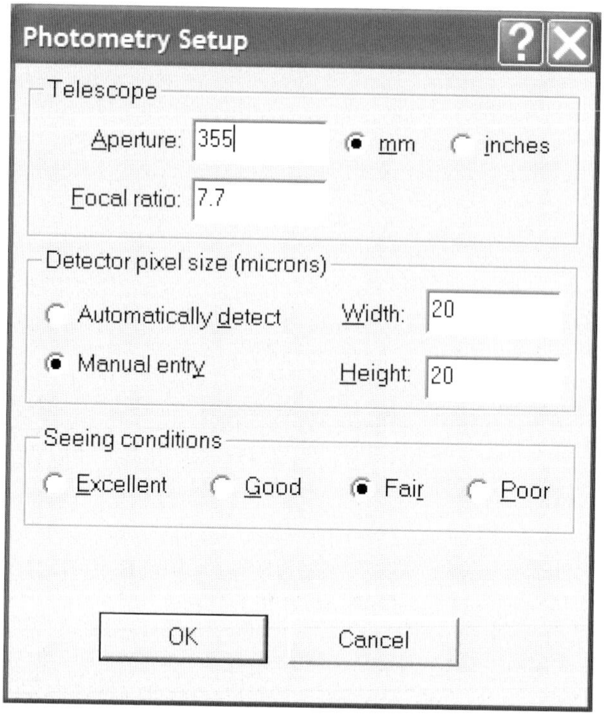

Figure 7.10. The Software Bisque *CCDSoft* photometry set-up window.

default Hubble Guide Star Catalogue, this is notoriously inaccurate as it was not intended as a photometric reference, merely as a source of guide stars. Therefore, despite the fact that *CCDSoft* can, when linked to *The Sky*, create a star chart for you, you may well prefer a chart produced for that particular field, with stars of photometric precision. However, this is often not possible when a bright supernova first appears. Unlike known variable stars, supernovae are brand new, and unless they occur in a bright Messier galaxy, an accurate chart may not be available for days or weeks after the discovery. In these situations, the best catalogue to use for photometry is the U.S. Naval Observatory's USNO UCAC2. The USNO CCD Astrograph Catalogue (UCAC) is an astrometric, observational program, which was started in 1998 and is just being completed at the time of writing. In fact, the sky has been covered to declination +40 or so for several years. It is just the final +40 to +90 catalogue that has taken the past few years to complete. The goal was to compile a precise star catalogue for fainter stars, extending the precise reference frame provided by the ESA Hipparcos and Tycho catalogues down to 16th mag. This is not only a very accurate astrometric catalogue, it is reasonably accurate for photometry, too. Unlike some of the larger catalogues, it will conveniently fit on a modern hard disk, too (taking up roughly 2 gigabytes). Other U.S. Naval Observatory catalogues, like the 80 gigabyte USNO-B1.0 can be accessed via the Web (http://vizier.u-strasbg.fr/viz-bin/VizieR/). So, to get back to the plot, *CCDSoft* can be used to measure star magnitudes, but it is far more powerful when combined with *The Sky*, which can actually create a star chart for you from the catalogue you are using, once you have set up all the required parameters and accessed the "Research/Comparison/Star Chart" menu. Full details are given in the *CCDSoft* manual of course. *CCDSoft/The Sky* can also be coaxed into producing a light curve from a set of images in a folder. The light curve data can then be text imported into Microsoft Excel's spreadsheet package.

AIP

The impressive book and software package *AIP*, which has now evolved into a second-generation product called *AIP4Win*, has become a firm favorite amongst many variable star photometrists. It has a slick photometry tool as well as a highly comprehensive photometry chapter that is well worth reading if you plan to become a supernova imager. The photometry tool is easily accessed under the "Measure-Photometry" menu and you are then presented with a choice of options entitled: single star, single image, multiple image, and extractive photometry (see Figure 7.11). In practice, the single image tool will probably be the most useful for the supernova photometrist. The great thing about the *AIP* photometry tool is that you can see the aperture and inner/outer annulus rings on the screen surrounding the stars you are interested in (see Figure 7.12). This helps greatly in

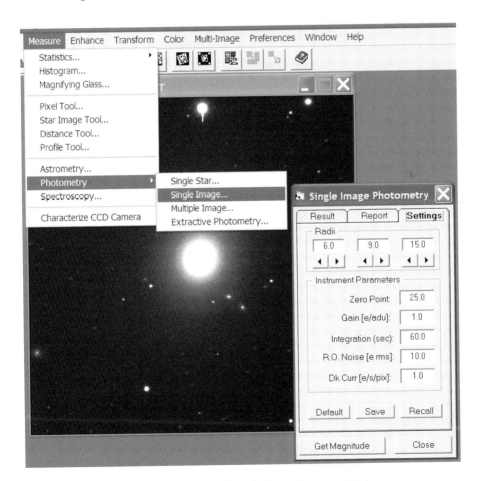

Figure 7.11. *AIP*'s Photometry options and Single Image Photometry Tool.

Figure 7.12. The *AIP* photometry tool shows the aperture and annulus rings for the target and comparison stars with great clarity.

choosing an appropriate aperture to surround the stars and an appropriate annulus for the background sky and to avoid any faint stars that might complicate the measurement. The first star you click on is the variable star (V) and the second star is the comparison star (C1). If you click on a second comparison star, then that can serve as extra security just in case the first comparison star turns out to be variable. The photometry tool has various controls which are accessed via three tabs labeled "Result", "Details", and "Settings". The "Result" tab, not surprisingly, displays the results from the measurements in various ways. The "Details" tab provides information on factors such as the number of pixels in the aperture and annulus, the signal-to-noise ratio, the number of comparison stars, and the standard deviation in the measurements. The "Settings" tab allows you to actually set the aperture and sky background radii as well as various instrumental factors that may be needed to enable the absolute magnitude value offered (e.g., 15.7) to appear in a sensible range. For example, the gain of the CCD camera can be entered at this point (expressed in electrons per ADU) as well as the exposure time, read-out noise, and dark current. As with any comprehensive scientific package, there is a lot to digest to get precise photometric results, but in terms of comprehensiveness and value for money, *AIP4Win* is hard to beat. It also has a feature that enables you to extract the photometry to an Excel spreadsheet. Within the BAA, their Variable Star Section has devised a useful Excel template for *AIP4Win*.

Astrometrica

The final package I would like to mention is Herbert Raab's *Astrometrica*, which, confusingly, is *not* a photometry package. Nevertheless, it is such a brilliant piece of shareware that can pin down the position of an object so easily (whether an asteroid supernova, variable star, or whatever) that the additional photometric information it offers is hard to ignore. *Astrometrica* can be downloaded from http://www.astrometrica.at/astrometrica.html. After the hundred-day trial period expires, you need to pay a trivial €25 fee to Herbert Raab, unless you are unbelievably miserly, in which case you download it again. *Astrometrica* could not be simpler to use, provided your image files are in the standard astronomical FITS (Flexible Image Transfer System) format or in an SBIG CCD camera format. You simply run the program and then, if you are using the UCAC2 catalogue, decide which declination region the object is in. This will break down into a region covered by one of four CDs that may be copied to your hard disk, that is, Disc 1 (−90 to −37), Disc 2 (−37 to +1), Disc 3 (+1 to +40), and Disc 4 (+40 to +90 and about to be completed as I type these words). Of course you may choose another U.S. Naval Observatory catalogue such as USNO-B1.0, by accessing the Web. To select the correct region, you simply point the "File/Settings/Environment/Star Catalogue" to your preferred catalogue. You then load your FITS/SBIG format image and make sure north is at the top. You can flip the image vertically and horizontally if it has south at the top or is mirror-imaged. If you then select "Astrometry/Data Reduction" and click "OK", *Astrometrica* will auto-identify the field (using the FITS header info or your overriding data) and wham, it identifies all the catalogue stars in the field. If you then mouse-click on the object of interest, it spews out the precise RA and Dec and the magnitude too, as a bonus. Despite being advertised as an astrometric tool, *Astrometrica* is quite a handy photometric tool, too. No, it doesn't have the power and versatility of *AIP4Win* but it is reliable, easy to use, and cheap; a rare combination indeed. In addition, its speed and power in recognizing the USNO stars in your CCD field is second to none. We will meet *Astrometrica* again when we discuss astrometric measurement of new discoveries.

Light Curves

A single visual or photometric magnitude measurement is of limited use to astronomers. It is only when a complete light curve is available that the behavior of a supernova can be defined. Most supernovae, especially bright ones, are spectroscopically imaged by professional astronomers within days of their discovery. The professionals have a huge advantage in this area. The light grasp of professional telescopes in the 2- to 10-m class dwarfs anything that even the largest amateur telescopes can achieve, and light is vital for producing spectra. Spectroscopy is also a very specialized field, as we shall see later, and while a few amateurs worldwide have obtained spectra of the very brightest supernovae, it is in producing light curves, not spectra, that amateurs are renowned. Obtaining an initial spectrum of a suspect is vital for determining whether an object really is a supernova, and of what type. But even within different types there are often surprises in store for the patient photometrist. Professional observatories often neglect following the fade of a supernova; they have deeper cosmological targets to concentrate on and there are just too many supernovae to image them all. However, amateur astronomers are very competent in this area. In terms of absolute magnitudes, we have already seen that Type Ia supernovae dazzle with the brightness of billions of suns when they peak at −19 or −20. Type II events more typically peak at absolute magnitudes of around −17. So, it is often possible to make an initial guesstimate of a supernova's type simply by the first magnitude estimates. If a supernova appears to put out as much light as the galaxy it lives in, there is a fair chance it is a Type Ia supernova, unless it happens to be in a very small or dusty galaxy. This is not meant as a hard and fast rule. Galaxies come in all shapes and sizes and are tilted to us at all sorts of angles. In addition, dust attenuates the potential light output of galaxies considerably. Nevertheless, those ultrabright supernovae are often Type Ia's and the brighter they are, the easier they are to spot. Fully one-third of the visual supernova discoverer Bob Evans' impressive tally of 45 supernovae are of Type Ia. In a galaxy like the Milky Way, one might expect 20 or so supernovae every thousand years, with only one in seven being Type Ia. For all supernovae, though, the spectra and light curves are the real decider as to supernova type.

At first glance, the light curves of different supernovae (Figure 7.13) may look very similar. There is a sharp rise to maximum brightness, a period at the maximum, and then a gradual decline. Many supernovae are poorly covered in decline and, even with occasional spectra being recorded, it is sometimes not possible to be 100% sure which subcategory a supernova fits into. In lists of supernovae one often sees just a basic classification of "Type II" when, in practice it may be a Type II-L or P or even a low-hydrogen Type IIb. Often it is only by careful amateur study that the real classification can be determined. It is significant that for the two best studied supernovae in recent decades (i.e., 1987A in the LMC and 1983J in M 81), neither object fit precisely into a category without extra mysteries having to be solved. 1987A and 1993J both showed a strange double-peak in their light curves.

Type Ia supernovae show the sharpest rise to maximum, and they often stay within a magnitude of maximum brightness for less than a month or so. Some 5 weeks (typically) after the maximum brightness occurs, the brightness may be three magnitudes below the peak. Then an abrupt change in the decline rate occurs and the supernova starts fading more slowly, typically a magnitude every 2 months. I should stress though that every supernova is slightly different, and I am just giving an indication of the average trend.

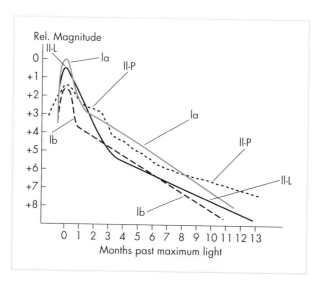

Figure 7.13. The typical light curve profiles for supernovae of Type Ia, Ib, II-P, and II-L.

Although less brilliant at their peak Type II-L supernovae show a similar light curve except that the rise to maximum brightness is slightly slower, and the fall from maximum is a bit more sedate, too. The peak in the light curve occurs at about the time that the temperature of the outer layers starts to decrease. The most noticeable difference between Type Ia's and Type II's is that the initial fading rate typically persists for some 10 weeks or more after maximum (i.e., twice as long as for Type Ia's). Ten weeks after maximum, a Type II-L supernova may be four magnitudes fainter than at its peak (that peak being, perhaps, only a tenth as bright as a Type Ia at the same distance). After that point, the slope changes to one of around a magnitude every 3 months, somewhat slower than for a Type Ia.

The light curve of a Type II-P supernova is rather more distinctive than either a Type Ia or a Type II-L. In this instance, there is a somewhat slower rise to maximum brightness and then a slow decline for a month or so, fading the supernova by just over a magnitude below the peak. A plateau phase then begins that may last a couple of months, before a brief week-long plunge by one magnitude. The plateau phase is thought to be linked to a change in transparency in the outer layer of the exploded star. As the shock wave propagates out through a Type II-P supernova, it heats up the outer envelope to more than 100,000 degrees Kelvin ionizing the hydrogen. Ionized hydrogen is less transparent and so light from the inner parts of the star cannot be seen, and so we can only observe photons from the outermost parts of the star emitted at a steady rate for the duration of the plateau. When this phase comes to an end, we see the week-long dramatic plunge, and this is followed by a more leisurely decline of around 1.5 magnitudes in 3 months. This decline rate slows by around 50% in the next 3-month period. The maximum brightness of Type II-L supernovae are typically about 2.5 magnitudes fainter than Type Ia supernovae, however, the peak brightnesses of Type II-P's show a large variation, probably due to differences in the original stars' diameters and masses. Big stars come in big, massive, and supermassive sizes.

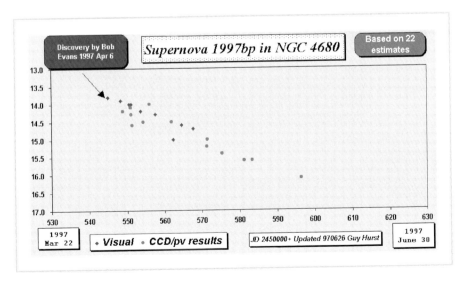

Figure 7.14. The light curve of Bob Evans' supernova 1997bp in NGC 4680, recorded by amateur astronomers as it faded from 14th to 16th magnitude. This was a Type Ia supernova observed in its initial fast decline phase. Courtesy: Guy Hurst/*The Astronomer*.

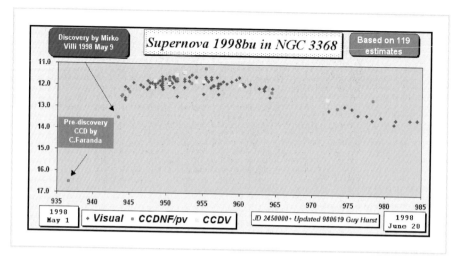

Figure 7.15. The light curve of Mirko Villi's supernova 1998bu in NGC 3368, recorded by amateur astronomers as it peaked at brighter than 12th magnitude and then faded to 13th. Note the Type Ia discovery occurred before maximum brightness and that a prediscovery image was found by Faranda a week before discovery. Courtesy: Guy Hurst/*The Astronomer.*

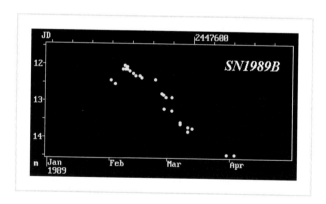

Figure 7.16. Another of Bob Evans' visual supernova discoveries, 1989b was well observed as it occurred in Messier 66. This was another Type Ia supernova that peaked at magnitude 12 and then characteristically declined by just over two magnitudes before its rate of decline slowed. Courtesy: British Astronomical Association Variable Star Section (BAA VSS).

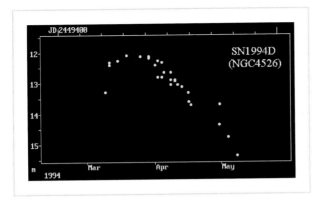

Figure 7.17. SN 1994D in NGC 4526 was discovered by professional astronomers Treffers, Filippenko, Van Dyk, and Richmond. A Hubble image of this supernova appears at the end of Chapter 1. This was another Type Ia standard candle. Amateur observations show how the supernova peaked at magnitude 12 and then declined fairly rapidly to 15th mag throughout March, April, and early May. Courtesy: British Astronomical Association Variable Star Section (BAA VSS).

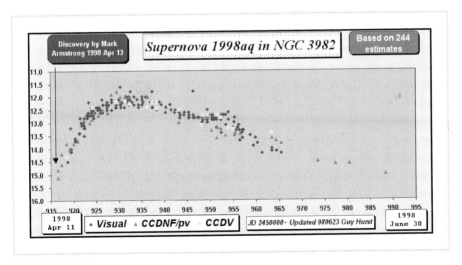

Figure 7.18. One of Mark Armstrong's discoveries, Type Ia supernova SN 1998aq in NGC 3982 was well observed throughout April, May, and June of that year. The light curve shows a rapid brightening after discovery, with maximum occurring 2 weeks later. A 6-week decline of just over two magnitudes ends with a change to a slower decline rate. Courtesy: Guy Hurst/*The Astronomer*.

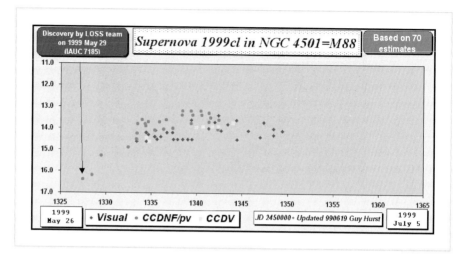

Figure 7.19. Another Type Ia discovery, this time by the Lick Observatory, and in Messier 88. This bright supernova's light curve was curtailed somewhat as the field entered U.K. summer twilight. Nevertheless, the classic 2-week climb to maximum and initial Type Ia decline were well covered. Courtesy: Guy Hurst/*The Astronomer.*

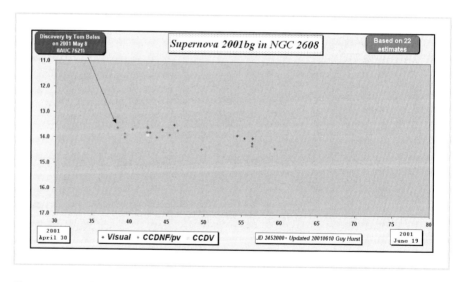

Figure 7.20. This Type Ia discovery by Tom Boles was monitored sparsely as the field, in Cancer, was sinking rapidly into summer twilight at discovery. Nevertheless, a steady decline was recorded. Courtesy: Guy Hurst/*The Astronomer.*

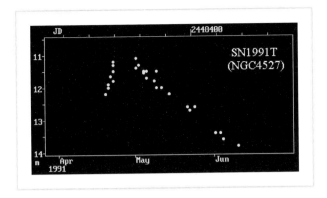

Figure 7.21. This very bright, and peculiar, Type Ia supernova in NGC 4527 was well observed in April and May 1991. It was discovered by several observers, namely, Knight, Bob Evans, Villi, Cortini, and Johnson. A photograph of this supernova, taken by the author, appears in Chapter 1. Professional photometry of this supernova extended for 3 years after maximum and showed a peculiar slowdown in the decline rate in the second year. However, in the first few months a characteristic rise to maximum (somewhat complicated by cloudy conditions for U.K. observers) preceded a standard three-magnitude rapid decline. Courtesy: British Astronomical Association Variable Star Section (BAA VSS).

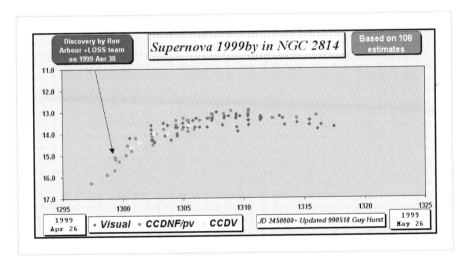

Figure 7.22. SN 1999by was another peculiar Type Ia supernova. This one was discovered by Ron Arbour and the Lick Observatory team; the fourth supernova in NGC 2841. A spectrum of this supernova, by Maurice Gavin, appears in Chapter 8. This supernova was *subluminous*, that is, fainter than would normally be expected for a supernova at that distance and possibly due to a smaller than normal nickel content. The limited light curve here shows a 10-day rise to maximum followed by the start of the decline phase. Courtesy: Guy Hurst/*The Astronomer*.

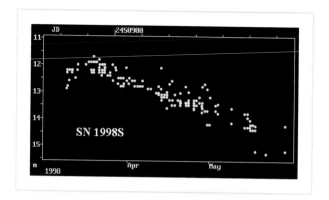

Figure 7.23. This bright supernova was discovered by the Beijing Astrophysical Observatory and shows the standard long decline characteristic of a Type II-L supernova. Courtesy: British Astronomical Association Variable Star Section (BAA VSS).

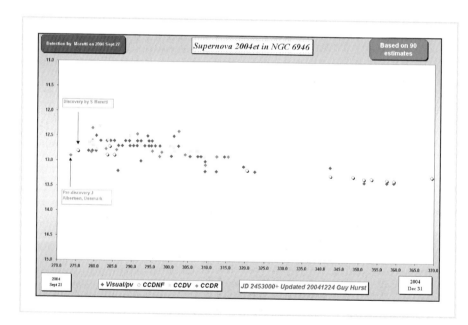

Figure 7.24. This is the eighth and most recent supernova in the superprolific galaxy NGC 6946, discovered by S. Moretti with an 0.4-m reflector. As can be seen from the very flat light curve, the supernova only faded by about a magnitude in 100 days, implying it was a very flat Type II-P. Courtesy: Guy Hurst/*The Astronomer*.

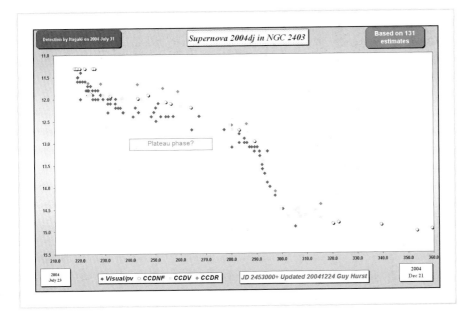

Figure 7.25. Supernova 2004dj provided Northern Hemisphere amateur astronomers with an almost unprecedented opportunity to observe a supernova in decline. Despite being discovered in July 2004, amateur astronomers were still measuring its brightness in early 2006, when it was still 17th magnitude. The light curve here shows the first 140 days after discovery. Clearly, there was an initial one magnitude decline, a plateau phase for 6 weeks, and then a rapid two magnitude decline, changing to a shallower decline—All characteristics of a Type II-P supernova. Courtesy: Guy Hurst/*The Astronomer*.

Figure 7.26. This is the famous and highly unusual double-peaked light curve of the bright supernova in Messier 81. The supernova was designated as Type IIb, that is, a core-collapse massive star, but with most, but not all, of its hydrogen removed by tidal stripping. A very unusual light curve indeed, as was the supernova 1987A in the Southern Hemisphere. Courtesy: British Astronomical Association Variable Star Section (BAA VSS).

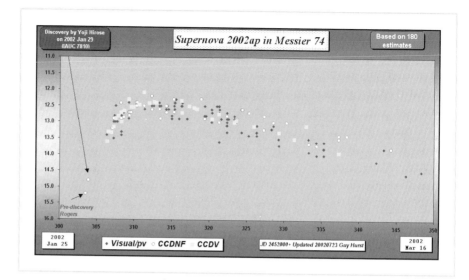

Figure 7.27. Another unusual supernova: SN2002ap in Messier 74. This object was discovered on January 29, 2002, by Japanese amateur astronomer Yoji Hirose at mag 14.5, and a week later it was brighter by two magnitudes. It was eventually classed as Type Ib/c and described as a hypernova, although one of the dimmer objects of this category, some of which can be 100 times brighter than Type Ia's! Cosmologists think the progenitor was probably a star equal to at least 40 solar masses! Courtesy: Guy Hurst/*The Astronomer.*

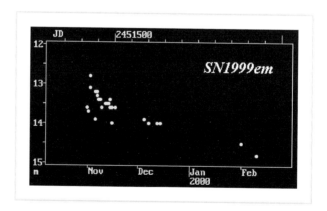

Figure 7.28. Supernova 1999em was discovered at the Lick Observatory. The light curve, although rather sparse, suggests it was of Type II-L. Courtesy: British Astronomical Association Variable Star Section (BAA VSS).

One year after maximum light, the brightness of Type Ia, Type II-L, and Type II-P supernovae will typically be about eight, eight, and six magnitudes, respectively below their peak magnitudes. The Type II-n supernovae that are characterized by narrow band emission lines on top of the broader emission features in their spectra tend to have a slow decline rate in the latter part of their light curves. For very bright supernovae such as SN 1980K in NGC 6946, SN 1993J in M 81, and SN 2004dj in NGC 2403, amateurs have been able to study their decline comprehensively over a period of 6 to 8 months after discovery with backyard equipment and a bit of perseverance. Type Ib supernovae tend to have a dimmer peak absolute magnitude than Type II-L's and, on average, slightly less than Type II-P's. These hydrogen-stripped Type Ib's also tend to fade at a faster rate than the Type II-L/II-P stars and at a rate more similar to their hydrogen-free smaller, binary system Type Ia cousins. Again, it is important to remind ourselves that Type I means hydrogen-free, Type II means hydrogen-present, and most supernovae are huge stars *except* the unique and valuable Type Ia's.

Figures 7.14 to 7.28 show the light curves of some bright supernovae that have been well-observed by amateur astronomers.

Chapter 8

Supernova Spectroscopy

Although a supernova's long-term light curve can be used to deduce its all important "Type" classification, astronomers like to classify them as soon as the discoveries are made. The only way to confirm quickly that a suspected supernova really is a supernova (and not an unusual variable star within our own galaxy, or a superbright nova in the host galaxy) is to record the spectrum of the star. Professional astronomers are often keen to establish whether the new object is a distant Type Ia at maximum so that they can accurately establish its implied distance. Unfortunately, even fairly large amateur telescopes will struggle to obtain a spectrum of an object that is fainter than you can see visually through the same instrument. This is because the light from the star is not focused to a point covering a few pixels, as a spectrograph splits the light from a star into a long, thin spectrum. If the spectrum is dispersed wide enough for useful resolution to be captured, it will be very faint, but a long CCD exposure can obviously help compensate for this. Even the simplest prism or grating (net curtains can be used!) will split the light of our own sun into its constituent colors and show dark absorption lines, produced by specific elements in the sun's atmosphere. The extreme visual limits of the solar spectrum stretch from the H and K lines of calcium at 3934 and 3968 angstroms, deep in the violet (Note: 1 angstrom is 10^{-10} m) to the hydrogen-alpha 6563 angstrom line, deep in the red. Of course with very distant galaxies, redshift can move these lines to much longer wavelengths. Any useful spectroscope needs to be able to record details across this range. Although astronomers can feel really starved of light when trying to get a supernova spectrum, there was one case when they had far too much light! When supernova 1987A went off in the Large Magellanic Cloud and rose from 4th to 2nd magnitude, professional astronomers suddenly found that the equipment on their best and biggest telescopes was not designed for getting a spectra or photometry of such a bright object. Indeed, the object was so bright compared with what they were used to that new instrumentation was quickly lashed together to exploit the high spectral resolution opportunity that SN 1987A made possible and also to attenuate the light from the supernova for photometry. As described in Paul Murdin's 1990 book *End in Fire*, the 3.9-m Anglo-Australian Telescope was effectively stopped down to 12-cm aperture, the South African Astronomical Observatory masked their 50-cm telescope to an effective 15-cm aperture, and astronomers at Cerro Tololo masked their 40-cm telescope until it had an equivalent aperture of 5 cm! Although it is very unlikely that the next, and long awaited, "galactic supernova" will explode into our nighttime skies in the next few years, it is not totally impossible. If it does happen, any amateur with a spectroscopic or photometric capability could be very well placed to obtain some very valuable spectra, while the professionals are still working out how to filter their equipment!

What Causes Spectral Lines?

Although most amateur astronomers have some appreciation of the fact that elements can be identified as vertical lines in a star's spectrum, a brief explanation of the origin of spectral lines may not be inappropriate at this point (see Figures 8.1 and 8.2). A bright, hot star emits radiation across the electromagnetic spectrum and, when an optical prism, a diffraction grating, or even raindrops are used to disperse the light into its component colors, we see the standard rainbow colors from red through to violet. However, when the light in question has passed through a gas on its way to us, photons may be absorbed by electrons orbiting the atomic nuclei in the gas and the electrons will then jump up a discrete orbit level as they absorb photons. This absorption leads to dark lines appearing at discrete wavelengths in the spectrum. If I want to be totally accurate here, I would have to admit that when photons are absorbed and an electron moves to a new energy level, the electron eventually returns to its original level, reemitting a photon. So you might think the effect would cancel out? In fact, because the photons can be reemitted in any direction but were absorbed while heading straight for us, there is a net dimming of the overall light (i.e., the dark absorption line).

The opposite effect occurs when the gas is being excited by some energy input. In this case, the electrons may jump down a discrete orbit level as they emit photons. This emission leads to bright lines appearing in the spectrum. The well-known Balmer series of hydrogen atom orbit transitions give rise to lines in the visible part of the spectrum and correspond with electron transitions between the second orbit level and higher orbit levels. It was only with the development of quantum physics and probabilities that the discrete allowable orbits were fully understood.

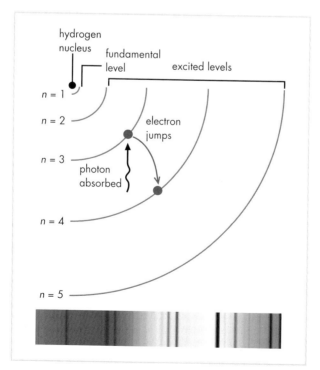

Figure 8.1. Photons passing through a gas may be absorbed by electrons orbiting the atomic nuclei in the gas, and the electrons then jump up a discrete orbit level. This leads to dark lines in the spectrum.

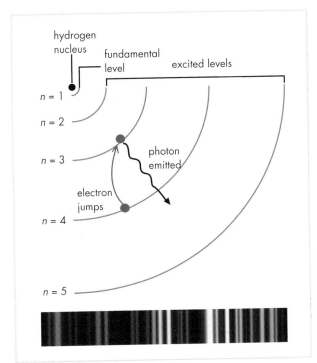

Figure 8.2. When a gas is being excited by energy, electrons may jump down a discrete orbit level as they emit photons. This leads to bright lines appearing in the spectrum.

How a Spectrograph Works

First, let us get some basic definitions, often encountered in spectroscopy, out of the way. A spectrograph is a device that can produce a graph of the intensity of light as a function of color or wavelength (i.e., a spectroscope that produces a graph). A spectrometer is a device that measures only one selectable color, whereas a monochromator is a device that transmits only one color. The basic components of a single prism spectrograph are shown in Figure 8.3. Essentially, the aim is to gather as much light as possible from the star being observed (and not from anything else), split the star's light up into a spectrum, and focus the spectrum. If a single, narrow beam of light from an intensely bright point-like source like the sun was being examined, all you would need is a chink of light and a prism. However, for astronomical spectroscopy with a telescope, where the star is much fainter (and there may be other stars nearby), you need to channel parallel light from the star through a prism and then use a lens to focus the red end of the spectrum at one end of the CCD detector chip and the blue end at the other. This is the simplest, most efficient and practical way to capture the spectrum. Moving from left to right in the figure, we first come to the slit. In a normal telescope, this is where the eyepiece would focus or the CCD would be placed, that is, the focal plane, where the image of the star field exists. The purpose of the slit is to reduce background noise from the rest of the sky and to reduce any overlap from adjacent wavelengths. The narrower the slit, the better the spectrum is resolved, *but*, if the slit is narrower than the focal plane star diameter, light will be lost. The collimator is simply a lens designed to ensure that parallel light enters the prism. Once the parallel light has been split into a spectrum by the prism, the spectrograph's own mini telescope

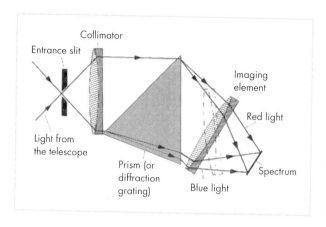

Figure 8.3. The basic components of a spectrograph. Diagram courtesy of Prof. Chris Kitchin.

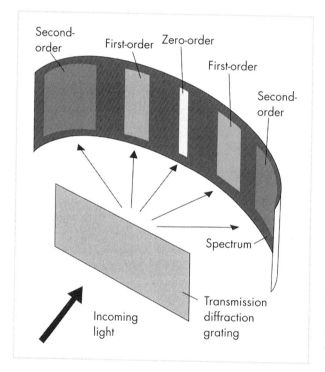

Figure 8.4. The production of multiple spectra of different orders by a diffraction grating. Diagram courtesy of Prof. Chris Kitchin.

lens, or imaging lens, easily focuses the red light to one end of the CCD and the blue to the other; so the spectrum is nicely spread out along the chip.

So, those are the basics dealt with. In practice, though, there are many variants. For example, the prism can be replaced with a diffraction grating, which disperses the light in a slightly different way. With a diffraction grating, dispersions are conveniently greater than with a single prism (older spectroscopes often used several prisms in sequence) but they produce two sets of spectra, each with several "orders" of spectra (see Figure 8.4). The majority of the light goes into the white light "zero order" spectrum. Thus, the spectra are not as bright. However, if the grating is of the "blazed" type (more often found in reflection gratings), the indi-

Figure 8.5. SBIG's auto-guiding spectrograph attached to an SBIG ST7 CCD camera. Light travels from A (telescope interface) via a slit to B (mirror) to C (collimating mirror) to D (the grating carousel). The spectrum produced is then directed to the second half of the collimating mirror (E), which then focuses it into the CCD camera (G) via another mirror (F). Photo: Courtesy Maurice Gavin.

vidual grating line surfaces are angled to direct the majority of the light into the spectrum. Obviously, to take advantage of this, the grating has to be angled accurately to direct the bright spectrum at the detector.

The next issues to be addressed are how well can the spectrum be resolved, what focal length should the spectrograph telescope lens be, and how much of the spectrum will fit onto the length of the CCD? With the typical prisms or gratings available to amateurs, the middle of the visual spectrum can be resolved as finely as 1 angstrom. But these same prisms or gratings typically disperse the spectrum such that 1 angstrom of the spectrum subtends an angle of, say, only 2 arc-seconds. Thus, the spectrograph's imaging lens will need a focal length of a meter to capture 1 Angstrom of resolution per 10 micron CCD pixel. At this scale, however, a 500-pixel-long CCD array will only capture 500 angstroms of visual spectrum, compared with the whole visual spectrum of 4,000 to 7,000 angstroms, that is, 3,000 angstroms. SBIG's Self-Guiding Spectrograph (Figure 8.5) features a choice of two dispersing diffraction gratings, offering 1 angstrom per pixel and 4.3 angstroms per pixel. But it also features an ingenious double concave mirror arrangement, which acts both as collimator and imaging lens and keeps the unit's size compact. However, for the homemade spectrograph builder who wants to keep the imaging lens focal length short, settling for a resolution of a few angstroms per pixel and using a CCD with small pixels will help keep the system compact. It is important not to get confused here between spectral resolution and dispersion. Let's look again at SBIG's Self-Guiding Spectrograph to clarify matters. The SBIG unit, like all spectroscopes, has a spectral resolution set by the diffraction grating's performance, but this can be compromised if the slit is widened (to reduce exposure times) and by instrumental deficiencies. However, to actually capture the resolution on the CCD, the dispersion and the focal length of the imaging lens/mirror must deliver a small enough "angstroms per pixel" scale. The SBIG Self-Guiding unit features a choice of two diffraction gratings of 150 and 600 lines per millimeter with corresponding resolutions of 10 and 2.4 angstroms with the narrow, 18-micron slit (18 microns = 2 arc-seconds at 2 m focal length). With the wide, 72-micron slit, the 150 and 600 line gratings deliver resolutions of 38 and 10 angstroms. The dispersions of these gratings, combined with the focal length of the imaging lens/mirror, give image scales of 4.3 angstroms per pixel with the 150 grating and 1 angstrom per pixel with the 600 grating. The image scale is always of finer resolution than the spectral resolution to ensure that all the resolution available from the instrument is captured at the CCD. Obviously, if the spectrum is analysed at a higher resolution on the CCD surface, the spectrum will be dimmer and longer exposures will be needed to capture a supernova's "Type" classification. In spectroscopy, everything is a trade-off between spectral resolution and brightness, but if your system reliably auto-guides on a star (like SBIG's SGS unit), long exposures can be combined with good resolution, and the crucial details can be resolved, even for supernovae that are below the visual threshold in the same instrument. Many amateur supernova discoveries are as faint as magnitude 17. With short exposures of a minute or two in length, it is virtually impossible for amateur spectrographs to get this faint on 0.3- or 0.4-m apertures. In practice, telescopes in the 1- to 2-m range are needed to routinely obtain such spectra with relatively short exposure times.

For the DIY spectroscope builder, optimum grating/prism assemblies are rarely available; likewise for the collimating and imaging lenses. It's usually a case of buying cheap components and bolting them together to see what happens. Amateur spectrographs are rarely designed precisely. Fortunately, diffraction gratings of 600 lines/mm can be purchased for as little as $25 and adjustable slits can be made from two razor blades. In addition, secondhand camera lenses can be called into service for the collimating and imaging lenses, leaving the CCD as the most expensive component. But there are other technical considerations, too. For example, how do you actually keep the telescope guided so that the star being analyzed is kept in the slit? One way of doing this is to focus a guiding eyepiece or telescope on the outer surface of the slit; this surface, if highly polished, will easily show the outer overspill of the star's disc. It is actually advantageous to let the star's right ascension drift trail back and forth along the slit length as this produces the height of the spectrum. With perfect tracking, the spectrum would be an almost infinitesimally thin line and very hard to analyze. The slit in the SBIG SGS unit is formed from two halves of a plane mirror, which reflects the image at the focal plane to the separate guiding CCD. Thus, while the main CCD collects the spectra, the guiding CCD shows the field, with a dark line (or white if back-illuminated) showing the position of the slit; perhaps the ultimate spectroscope luxury.

Few amateurs will want to spend $5,000 on the auto-guiding SBIG spectrograph, but, fortunately, the basic components of a spectrograph are easily available and just need loads of experimentation and patience to fine tune.

SBIG's SGS (Self-Guiding Spectrograph)

Without a doubt, the most exciting instrument in this field is the one I have already referred to, namely Santa Barbara Instruments Group's (SBIG) $5,000 SGS. SBIG now makes a less expensive instrument, too, described in detail below. SBIG's spectrograph is designed exclusively for use with their auto-guiding ST7 and ST8 CCD cameras, and with modern CCDs being three magnitudes more sensitive than the best spectroscopic photographic film, it can be seen that a modern 30-cm Schmidt-Cassegrain plus CCD will easily outperform a historic 1-m instrument using the spectroscopic films of the 1980s such as Kodak's 103 series. To all intents and purposes, astronomical photography is now resigned to history. Modern CCD results can be analyzed as soon as the exposure ends, and there is no horrible developing, fixing, and drying phase that I so well recall from my own early amateur astronomy days. So what science can be done with an instrument like the SBIG SGS? SBIG's own promotional literature quotes a healthy spectral signal to noise ratio of 10:1 for a 9th mag star with a 20-minute exposure using a non-ABG ST-7 CCD and a 25-cm aperture telescope in high spectral resolution mode (as good as 2.4 angstroms). The low spectral resolution mode (10 angstroms at best) will achieve the same signal-to-noise ratio with a mag 10.5 star using SBIG's narrow slit option. In practice, this specification tells us that spectra of bright galaxies (e.g., the Messier galaxies) taken with the SBIG unit can easily show the redshift due to the expansion of the Universe, when compared with the spectrum of a nearby star. Perhaps the most useful application of such an instrument in amateur hands is for the spectral monitoring of novae, which are usually discovered at mag 10 or brighter. Determining the spectral type of bright supernovae is also a possibility, although, even with CCDs, large amateur telescopes and long exposures are required. In theory, a large (0.4 to 0.5 m) aperture amateur telescope should be able to take a useful spectrum of a mag 15 supernova with an exposure of 1 hour with an instrument like SBIG's SGS; the signal-to-noise ratio would be poor, but good enough to discern the difference between, say, a Type Ia and Type II supernova. However, few amateurs seem to be doing regular supernova spectra work at the time of writing. More information on SBIG's products can be found at http://www.sbig.com/index.htm.

SBIG's Deep Space Spectrograph (DSS-7)

At 30% of the cost of SBIG's SGS, the DSS-7 will be a much more attractive proposition to many budding amateur spectroscopists. However, as with anything that is a cheaper version of the same product, some capability is lost. First, there is no provision for auto-guiding. The length of exposure will be limited by your telescope's tracking accuracy unless you have another CCD camera set up to work as an auto-guider or unless you guide visually with a piggyback telescope/off-axis guider. For stellar work, it may well be tricky to keep a star on the narrowest slit, unless you have a Paramount ME or similar quality mount that can track perfectly for several minutes. Having said this, a small tracking error in spectroscopic work is nowhere near as fatal as with "pretty picture" deep sky work. Indeed if it is only a few arc-seconds, it can be arranged so that it merely increases the height of the spectrum. It is only when the target leaves the slit that a real disaster has occurred. For nebulous objects, which are better suited to this cheaper instrument, the task is much easier because a bit of tracking drift will have a negligible effect. With a 35-cm telescope tracking perfectly for 120 seconds (not impossible with a quality mount), a very reasonable spectrum of a 12th magnitude star is possible, although this does limit the user to the very brightest supernovae. Putting the star in one of the wider slits of the DSS-7 helps (50, 100, and 200 micron slits are available) but there will be some blurring of the spectrum. The DSS-7 is optimized for use with SBIG's ST-7XME or the low-cost ST-402 and will work well with ST-8/9/10/2000 cameras and ST-237s. But it will not work with SBIG's STL series because of the built-in filter-wheel, and its extra back-focus requirement. Used with the 9-micron pixels of an ST-7XME, a resolution of 15 angstroms is possible. Maybe the key to using such an instrument to its limit is simply going back to the practices of the days of film, when astronomers were "real men" and would happily sit for hours at a guiding eyepiece with the star locked onto the crosshairs, whatever the outside temperature.

In the United Kingdom, Maurice Gavin has been the leading pioneer of CCD spectroscopy for many years and has obtained numerous spectra of unusual variable stars and novae using a 30-cm Meade LX200 and homemade spectroscopes. For a few very bright (mag 12 or 13) supernovae, Maurice has attempted to determine the spectral type of the supernova. Figure 8.6 shows his 15-minute exposure

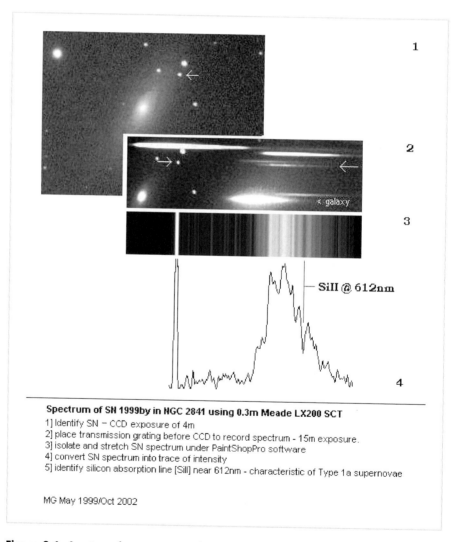

Spectrum of SN 1999by in NGC 2841 using 0.3m Meade LX200 SCT

1] Identify SN – CCD exposure of 4m
2] place transmission grating before CCD to record spectrum - 15m exposure.
3] isolate and stretch SN spectrum under PaintShopPro software
4] convert SN spectrum into trace of intensity
5] identify silicon absorption line [SiII] near 612nm - characteristic of Type 1a supernovae

MG May 1999/Oct 2002

Figure 8.6. Spectrum of supernova 1999by in NGC 2841 by Maurice Gavin: a 4-minute conventional image and a 15-minute spectrograph with a 0.3-m LX200 and homemade spectroscopic equipment. The silicon II absorption line at 612 nm is clearly visible, identifying the supernova type as Ia. Photo: Courtesy Maurice Gavin.

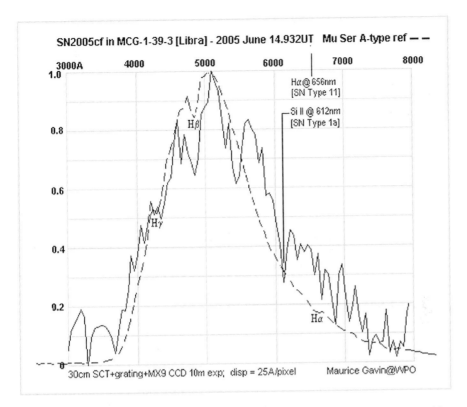

Figure 8.7. The spectrum of supernova SN 2005cf in the galaxy MCG 1-39-3, captured by Maurice Gavin with his homemade spectrograph. Again, the silicon absorption line clearly identifies this as a Type Ia supernova. The spectrum of the star MU Serpentis (dotted) is shown as a reference.

with a homemade spectrograph of supernova 1999by in NGC 2841. The silicon II absorption line at 615 nm is clearly visible, identifying the supernova type as Ia, a remarkable achievement. Figure 8.7 shows a similar result for supernova 2005cf in MCG 1-39-3. Figure 8.8 shows another spectrum by Maurice, this time showing the telltale hydrogen-alpha line in a Type II supernova.

Spectral Features of Supernovae

We have already seen that the essential difference between Type I and Type II supernovae is that Type I's have no hydrogen in their spectra, whereas Type II's do. Table 8.1 shows the key characteristics seen in the spectra of the various supernovae in the modern classification system.

Referring to the characteristics mentioned in the table: P-Cygni profiles are characterized by strong emission lines accompanied by corresponding blueshifted absorption lines. They are produced by an expanding envelope of gas. Most of the gas (that is not traveling in our direction) is traveling roughly perpendicular to our line of sight, producing emission lines that are not Doppler shifted. But the gas

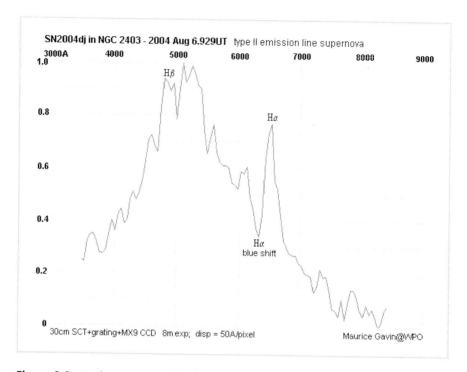

Figure 8.8. Another spectrograph by Maurice Gavin, this time for the very bright Type II supernova, SN 2004dj, in NGC 2403. The Type II classification can be deduced from the presence of hydrogen in the spectrum and the lack of a silicon absorption line.

Table 8.1. Spectral Characteristics of Supernovae

SN Type	Spectral Characteristics (Type I's are all Hydrogen Free)
Ia	6150 angstrom Silicon absorption at peak; iron emission lines seen later
Ib	~5700 angstrom (helium) absorption; oxygen and calcium emission lines seen later
Ic	No helium absorption; oxygen and calcium emission lines seen later
II-L	Low hydrogen content in spectrum or the P-Cygni profile is missing
II-P	Healthy hydrogen content in spectrum and a P-Cygni profile
II-n	Hydrogen in spectrum plus narrow emission lines and broad emission features
II-b	Low hydrogen/significant helium; oxygen, calcium, and hydrogen emission lines later

that is traveling toward us is seen directly in front of the star, and this produces the blueshifted absorption features in the star's continuum. P-Cygni profiles are especially prominent in Type II-P supernovae, which are assumed to have massive red supergiant progenitor stars.

When trying to identify spectral lines in amateur spectra, there are two pieces of software that are very useful. First, there is SBIG's own software, SPECTRA, and, second, there is an excellent freeware package called VSpec, available from Valerie Desnoux at http://astrosurf.com/vdesnoux/. This latter package is Windows based and used by virtually everyone in the amateur spectroscopy field. The professional package IRAF can be used and is very powerful. However, most amateurs will avoid a Unix/Linux-based product with a steep learning curve, especially if something far more friendly, like VSpec, is available. Some amateurs use an emission lamp, like a thorium/argon lamp, to precisely calibrate their spectroscopes, but this is far from essential.

Chapter 9

Amateur Supernova Hunting in the 21st Century

The first amateur supernova discoverer's name is probably unknown even amongst many top amateurs today. His name was Giuliano Romano of Treviso, Italy. He discovered two supernovae photographically using a 40-cm Newtonian reflector and later went on to become a professional astronomer. His discoveries were of 1957B in M 84 (NGC 4374) on April 23 of that year (3 days before the first ever BBC *Sky at Night* program by Patrick Moore) and 1961H in NGC 4564 on May 2 of that year. Both discoveries were very bright ones. 1957B was discovered at magnitude 12.5, and 1961H was discovered at mag 11.2. Indeed 1961H was one of the brightest supernova discoveries ever made (number 10 in my list).

The third supernova discovery by an amateur astronomer was made on July 17, 1968, by the experienced observer Jack Bennett of Pretoria, South Africa. Bennett discovered supernova 1968L visually while searching for comets with his 5-inch (12.7-cm) comet-sweeping refractor. The new star was discovered in the prolific supernova producing galaxy M 83 (NGC 5236), just 5 arc-seconds west of the nucleus. SN 1968L was discovered at magnitude 11.9 and was of Type II. Just over a year later, Bennett became world famous for his discovery of comet Bennett 1969i, one of the best comets of the 20th century. (Bennett discovered a second comet in November 1974.)

Another 11 years would elapse before the fourth amateur supernova discovery. Supernova 1979C was also discovered visually, this time by Gus Johnson using a 20-cm telescope. SN 1979C was discovered in M 100 (NGC 4321) on April 19 of that year. The discovery magnitude was 12.1 but it was discovered some 2 weeks after it initially exploded (probably around April 4). The supernova was of Type II-L and the new object was a whopping 56 arc-seconds east and 87 arc-seconds south of the nucleus.

From 1981 to 1997, one man would dominate the world of amateur supernova discovery: the Rev. Robert Evans of New South Wales. By April 1997, Evans had discovered an incredible 38 supernovae visually using mainly a 25-cm Newtonian (a larger, more cumbersome, 40-cm Newtonian was also employed at times for the later discoveries). A few other amateur astronomers did succeed in those 16 years (Okazaki, Horiguchi, Kushida, Johnson, and Aoki, for example), but it was only when U.S. amateur Michael Schwartz used an early robotic Paramount mounting mated to a Celestron 14, in 1997, that Evans' rate of discovery was under threat.

Both Schwartz, at his Tenagra Observatory, and a second U.S. amateur, Tim Puckett, were systematically discovering supernovae in large numbers from 1998 onwards. Around the same time, the U.K. amateurs Mark Armstrong (from 1996) and Tom Boles (from 1997), using humbler equipment at first (25-cm LX200s), embarked on a supernova discovery quest themselves. By 1999, with all four observers owning super-reliable Paramount mountings, the combined discoveries of Schwartz, Puckett, Armstrong, and Boles meant that the era of total Evans domination was over. However, Evans is still regarded with the greatest awe as his discoveries were all made visually and by memorizing hundreds of galaxy fields. Surely he is still the most incredible supernova discoverer in the world.

At the time of writing (September, 2006) these top six amateurs names were associated with the following remarkable numbers of discoveries:

Michael Schwartz: 33 + a further 258 discoveries as part of the Lick/Tenagra (LOTOSS) collaboration

Tim Puckett (and collaborators): 140

Tom Boles (all discovered while working alone and without collaborators): 103

Mark Armstrong (all discovered while working alone and without collaborators): 73

The Rev. Bob Evans: 46

Berto Monard (all discovered while working alone and without collaborators): 46

As 2005 came to a close, Tim Puckett overtook the legendary professional astronomer Fritz Zwicky's 123 supernova discoveries, a record that for many years looked like it would never be beaten.

Rival Systems

Sadly, amateur supernova hunters not only have their fellow amateurs to compete against when trying to hunt down supernovae. During the 1990s and the early years of the 21st century, a whole army of remorselessly efficient professional patrol telescopes have stolen the vast majority of supernovae from amateur hands. In many ways, this professional advantage has always been the case. Unlike with comets, amateur astronomers have never held the advantage with supernova discoveries. The vast majority of supernovae are just too faint for amateurs to dominate the scene. Even with CCDs now in amateur hands, the professionals trawl in hundreds of sub-magnitude 20 discoveries, whereas half a century ago the huge professional Schmidt cameras collected almost all of the discoveries in the magnitude range that 21st century amateurs now patrol. At the time of writing, the Pan-Starrs project, which will image 6,000 square degrees per night, to magnitude 24, in the search for near earth asteroids, looks like another serious threat. In 2005, the Sloan Digital Sky Survey (SDSS) was raking in the most supernova discoveries per year (130), but mainly in the magnitude 20 to 23 range (i.e., fainter than most amateurs are searching). Almost all amateur supernova discoveries are brighter than magnitude 19.0. The biggest threat to amateur supernova hunters in 2006 is still the Lick Observatory Supernova Search (LOSS), which in 2005 raked in 82 supernova discoveries, mainly in the magnitude range of 17 to 19 but with quite a few 16th magnitude catches, too. With CCD equipment, the vast bulk of amateur supernova discoveries are in the magnitude 15 to 18 range. Amateur discoveries fainter than magnitude 18 are fairly rare for obvious reasons. You need a telescope of at least 35 cm in aperture, perfectly focused, and in excellent conditions to detect one, and it must be reasonably elongated from the galaxy nucleus or an 18th magnitude suspect will be swamped in the galactic bulge. Supernovae brighter than magnitude 15 are rare and although they are obvious when imaged, the chances of bagging one more frequently than once or twice a year are slim even for dedicated patrollers. So any professional patrol that has regular successes in the mag 15 to 18 range is an enemy to the amateur supernova hunter. Fortunately, there is still hope though, not least because the dedicated high-Z professional patrols (i.e., those looking for high redshift supernovae) are tending to trawl through thousands of distant galaxies in clusters, but not specifically targeting the bright, high-prestige Messier and Caldwell galaxies. Also, there is still no foolproof system for automatically discovering supernovae. The traditional software systems used by professionals rely on image subtraction to subtract a master reference image from a new patrol image, after every image has been automatically subjected to a hot pixel/cosmic ray removal routine. Anything left over after the artifact removal and subtraction routines triggers an alarm notifying a team of human observers (typically undergraduate students). However, varying night-sky transparency, the presence of moonlight, and twilight or drifting cloud can confuse the software. In most automated patrols, the intervention of a human observer is still required at some point. The software task is especially complex for supernovae because they invariably sit on top of a fuzzy galaxy and, unlike a new asteroid, they rarely sit on top of a black sky. The human eye and brain can still make a more astute judgment when a supernova is only slightly brighter than the galaxy background. It goes without saying that any supernova patrol system has to guard against nearby asteroids or variable stars being confused for new discoveries. Although the Near Earth Asteroid Tracking (NEAT) telescopes have discovered more than 80 supernovae prior to 2004 (but none in 2004 or 2005), these telescopes are principally searching for moving objects. Indeed, the prolific LINEAR (Lincoln Laboratories Near Earth Asteroid Research) telescopes have not discovered any supernovae because their algorithms specifically search for fast-moving objects crossing the sky and not just for differences between a master and a reference image. So, despite a whole host of frighteningly efficient professional patrols, the amateur supernova hunter still has a role to play.

A Numbers Game

We have already seen that there is some dispute about how often supernovae occur in galaxies. There are so many factors to bear in mind that trying to calculate the probability of success based on predicted supernova occurrence rates is just a lottery. Galaxies can be big or small and they can contain varying amounts of young and old stars. They can also contain various amounts of obscuring dust and can be tilted toward us so they are edge-on or even face-on (and any angle in-between). A number of superprolific galaxies can raise ones expectation of success. For example, the two most prolific supernova producing galaxies are NGC 6946 and NGC 5236 (M 83). Since 1917, eight supernovae have been found in NGC 6946, namely in 1917, 1939, 1948, 1968, 1969, 1980, 2002, and 2004. Since 1923, six supernovae have been found in M 83, namely in 1923, 1945, 1950, 1957, 1968, and 1983. Both galaxies are large spirals seen almost face-on from Earth. At face value (pun intended), their productivity would appear to indicate that such galaxies can produce detectable supernovae every decade: almost four times the amount that such galaxies might be expected to produce (let alone be discovered). However, all galaxies are different, and by picking these two we really are picking exceptional cases. A naïve patroller might imagine that by patrolling, say, one hundred such galaxies, he or she could bag a supernova every 5 weeks or so. Dream on! It is now more than 120 years since the last supernova was seen in the Andromeda Galaxy, M 31, and more than 400 years since the last seen in our own Milky Way Galaxy.

To find out how many galaxies need to be patrolled for regular discoveries, one simply has to look at the statistics of the successful patrollers. Take the British supernova patroller Mark Armstrong for example. Mark currently has 73 supernova discoveries to his credit. From 1994 to 2003, his peak patrol years, Mark patrolled constantly, first with a 25-cm LX200 and, by the final years, with three ultrareliable Paramount/Celestron 14 systems. In that 9 years he checked a third of a million galaxy images on 1,100 nights for 60 discoveries. That is an average of one discovery for every 5,500 patrol images. In 2002, with three Paramounts patrolling, he observed on 109 clear or partly clear nights and exposed 83,385 galaxy images: that's an average of 765 images per clear night. The two newer Paramount ME systems really delivered the goods in that year, bagging one discovery for every 2,900 images. His countryman Tom Boles has reported similar success rates from the same, mainly cloudy, British skies.

The Katzmann Automatic Imaging Telescope (KAIT) at the Lick Observatory (see Figures 9.1 and 9.2) images up to 1,250 galaxies per night (30-second expo-

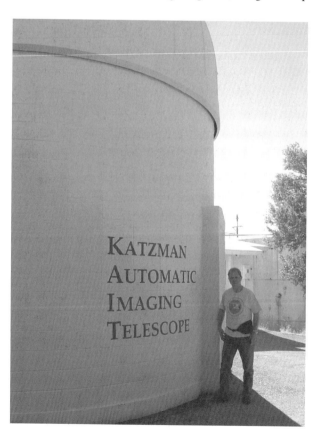

Figure 9.1. U.K. supernova hunter John Fletcher stands beside the dome of the Katzmann Automatic Imaging Telescope (KAIT) at Lick Observatory atop Mount Hamilton, just east of San Jose, California. This is the world's most successful supernova discovery telescope.

Figure 9.2. The world's most productive professional supernova patrolling telescope, the KAIT (Katzman Automatic Imaging Telescope), and principal investigators Weidong Li and Alex Filippenko. Photo: Courtesy Weidong Li/Lick Observatory Supernova Search.

sures with a 0.76-m aperture, reaching mag 19.5) on 300 clear nights per year and, in recent years, has discovered 80 to 90 supernovae per year. So, again, an average of one discovery for every 3,000 or 4,000 images seems to be the norm there, too. From 1998 to 2005, KAIT discovered more than 500 supernovae making it the world's most successful supernova discovery telescope. The total number of galaxies in the LOSS database is 14,000 but, of course, that whole database is not available in the night sky all year round. Depending on the time of year, and the position of the moon in the sky, in the past LOSS has revisited galaxies as regularly as every 2 days, or as infrequently as every 10 days. Nevertheless, this is a far better frequency than most cloud-bound amateurs with a day job and family commitments could ever hope to achieve. The KAIT facility, located at Lick Observatory, just east of San Jose, California, is led by Prof. Alex Fillipenko and Dr. Weidong Li, as well as a team of half a dozen research students who help with checking the images. KAIT was not the first automated supernova discovery project. Dr. Sterling Colgate received much publicity in the 1980s when developing a prototype 75-cm f/6 telescope for the same purpose. However, Colgate's telescope never delivered its hoped-for performance, whereas KAIT, spectacularly, has.

Supernova patrolling is a numbers game. Of course, if there were no other competing systems, a lone patroller would have a much better discovery rate. From most countries, though, the majority of nights are cloudy and it is during these cloudy periods that the competing professional systems (especially LOSS) steal the supernovae as soon as they appear on their images. Although there are hods of galaxies in the night sky, especially in the Northern Hemisphere spring sky, amateur patrollers and LOSS are often covering the same database of bright galaxies. This overlapping database (for the whole year) typically consists of 10,000 or 11,000 galaxies if one includes the faintest galaxies it is practical for amateurs to image. Recently, with competition from more powerful large aperture systems, the LOSS team have reduced their database of galaxies so they can patrol the brighter ones more often and catch supernovae on the rise. With the potential to image more than 1,200 galaxies per night, if a database of 1,200 is employed every galaxy can be checked every night and supernovae can be caught on the rise to maximum far more often. Even though an amateur astronomer's CCD fields are small, typically between 10 and 15 arc-minutes, it is often possible to squeeze two or even three galaxies on one image. The Abell and Hickson catalogues, plus a recent amateur Atlas of Galaxy Trios (see Appendix), can be useful for planning this sort of strategy. Planetarium software like *Guide 8.0* will show you where the brighter clusters are with a few keypresses.

A Realistic Strategy

You don't have to be a genius to work out that to discover supernovae you have to be nothing short of obsessed! A fascination with supernovae needs to be in your blood, as well as a highly competitive nature. Supernova hunting is about science and also about competition with professionals and fellow amateurs. As yet, no automated system has put the amateur astronomer out of business in this field, because spotting something that has changed in appearance is built into the human brain. We can all recognize thousands of faces of friends, politicians, and TV or film stars. If a friend or relative changes in appearance, we instantly spot that something is not quite right. Pattern recognition is second nature to human beings. We can spot something different from various angles and under various different illuminations. Even today, this is a horrendously difficult task to achieve with automated software. If cloud partly covers a galaxy, or if conditions are hazy (or in twilight or out of focus), we instantly know what has happened. Only the most sophisticated software can work this well, and even then, not *quite* as well. In fact, where faint supernovae are only just visible above the general glow of a galactic nucleus, human checkers still reign supreme. However, professional software has one overwhelming advantage. It never tires of checking and it never loses the will to live with the tedium of it all. If you use powerful software to flag up the most likely supernova candidates and a team of university students to sift through these candidates, you combine the tireless efficiency of software with the judgment and perception of the human brain: a potent combination. That is how the professionals do it.

The vast majority of successful supernova hunters that I know have no daytime job and no children either. Early retirement, wealth, or a (temporarily!) very tolerant spouse have enabled them to dedicate their lives to supernova patrolling. Supernova patrolling is not something that can be taken lightly. You have to patrol thousands of galaxies on a regular basis or you will not succeed. In addition, the images have to be checked as soon as they are taken because competition is fierce. However, there is no reason why a casual supernova patrol cannot be undertaken while obtaining "pretty pictures" of some of the brighter galaxies. Such pictures can be useful to other amateurs if, for example, an unusual variable star is located in the field, or, to eliminate a very faint supernova that is too close to the limit of the potential discoverer's master image. Carrying out a casual, but sustainable, supernova patrol is a bit like taking part in the national lottery, but with the added bonus of building up a nice library of pretty galaxy images. Some of the more detailed Messier and Caldwell galaxies can look stunning just on the initial raw image.

Hardware for the Task

I will deal with the options for the visual supernova patroller shortly, but the vast majority of successful supernova patrollers use hi-tech robotic equipment so I will deal with the hardware requirements for this first.

We have already seen that the prime targets for the amateur supernova hunter, using CCD equipment, lie in the brightness range of magnitude 15 to 18. We have also seen that, on average, thousands of galaxies will need checking to get at least one discovery. To discover supernovae at the fainter end of this scale, and at the rate required, we need, realistically, to be able to image down to magnitude 19 in 60 seconds or so. Trying to discover supernovae on the very limit of our equipment will just lead to false alarms. So, to be a successful patroller, a reliable "GO TO" telescope system with an aperture of at least 25 cm is required. A consideration of image scale is vital, too. Michael Schwartz' original Celestron 14/Paramount system, and that of Mark Armstrong and Tom Boles, employed cameras with 20- to 24-micron pixels at the full f/11 focal length of a C14. This gave an image scale of 1 arc-second per pixel, but a field of view of 10 arc-minutes or less. The U.K. supernova patroller Ron Arbour has successfully employed a 30-cm Schmidt-Cassegrain for patrolling and, at the time of writing, has bagged 16 supernovae with image scales closer to 2 arc-seconds per pixel (using Starlight Xpress MX916/SXV H9 cameras at f/6.3 and f/5). Typically, patrollers working at 1 arc-second per pixel can only fit one galaxy at a time in each field, although, in galaxy-rich areas two or more are possible. However, with bigger mega-pixel CCDs now becoming commonplace, I do wonder if farming galaxies with wider field CCD systems is something, as yet, largely unexploited by the amateur patroller? The recent liking for very long focal length systems largely stems from the success of Michael Schwartz with his original Celestron 14 working at f/11. Undoubtedly, an image scale of 1 arc-second per pixel really helps in weeding out faint supernovae close to galactic nuclei, but then so does a sharp focus, good collimation, and not overexposing the nuclear regions. Also, the numerous Paramount-based systems (invariably employing Celestron 14s) are capable of slewing to arc-minute accuracy so that 10 arc-minute fields of view are not impractical, with regard to hitting the target time after time. In the late 1990s when long focal length patrolling with large pixel (20 to 24 micron) CCDs became popular, cooled mega-pixel systems were prohibitively expensive. However, now that digital SLRs with 36×24 mm sensors are becoming commonplace, and filtering down to the astronomy market, farming numerous galaxies in one shot is practical. Let us look at a couple of possibilities.

SBIG (Santa Barbara Instruments Group) currently market their STL 11000 series CCD cameras with 36×24.7 mm sensors in a $4,008 \times 2,672$ array and 9-micron pixels. Imagine a 200-mm aperture f/4 Newtonian, with a coma corrector/flat-field corrector, using such a system. The field of view would be 2.6×1.8 degrees with an image scale of 2.3 arc-seconds per pixel. With a 200-mm aperture, discovering supernovae as faint as magnitude 17 would be achievable with 60-second exposures and the field of view area would be 170 times larger than with a 10 arc-minute wide system. Surely such a system would be ideal for areas densely packed with galaxies, such as Virgo and Coma Berenices? Okay, you would not necessarily get 170 galaxies in one shot, but you could certainly collect 40 galaxies brighter than magnitude 15 in one shot. This sounds like quite an impressive system. Alternatively, you could opt for a more affordable Starlight Xpress SXV M25 system with 7.8-micron pixels in a $3,024 \times 2,016$ grid (covering 23.4×15.6 mm). With the same 200-mm aperture f/4 system you would get 2.0 arc-seconds per pixel and a field of view of 1.7×1.1 degrees. Of course, by the time you read this book, new CCD toys will be available, at a lower price and, maybe, with even bigger CCDs. Wide-field systems are of little use though if aberrations creep in at the field edge.

More expensive optical systems, such as Ritchey-Crétiens, have longer focal ratios (typically f/7 or 8) but are sharp over a much wider area than a Schmidt-Cassegrain of the same size. Mounting the optical tube assembly of something like Meade's RCX 400 telescopes on a reliable Paramount ME mounting, using a big CCD chip, might make for a good system that would cover a relatively large field combined with superb tracking, pointing, and reliability.

With really wide fields of view, your mount needs a considerably less accurate slewing capability, too. Indeed, with a field of view 2 degrees across, there is no reason why an observer could not conduct a supernova patrol using mechanical setting circles to reset the position. However, an automated system is highly desirable as it never gets tired and the observer can watch the images downloading and give them a quick visual scan before the next image downloads. The GO TO systems available with standard Schmidt-Cassegrain's of 20-cm aperture and larger can slew to an accuracy of about ±5 arc-minutes over quite large distances from a calibration star, provided the mount is accurately polar aligned. (Yes, I know the manufacturer's may claim finer pointing accuracy, but I am being realistic!) Thus, with such systems, a field of view of 15 arc-minutes and more is highly desirable. Experienced amateurs have managed to improve the pointing accuracy of such telescopes by adjusting the worm and wheel and addressing the flexure problems with such systems, but, in practice, a 5 arc-minute error is fairly typical with non-Paramount systems when slewing tens of degrees. Indeed, in the worst examples, errors of several arc-minutes can result solely from the SCT (Schmidt Cassegrain Telescope) mirror tilting (sometimes called flopping) as it moves around the sky. Software is available to predict some of these sources of error, and correct them (see next section) but many amateurs will prefer to opt for the solution of a wider field of view.

Robotic Remote Control

One of the most powerful advantages of a GO TO system is that it is invariably possible to control a telescope remotely, so the observer is not additionally fatigued by the cold, the damp, a bombardment of suicidal moths, and stressed by his or her PC getting covered in dew. This advantage is colossal when supernova patrolling. If downloading galaxy images can be supervised in a comfortable, warm room, the endurance of the observer increases 10-fold, at a stroke. To control any telescope remotely, three items need to be under the observer's control, namely, the telescope, the CCD camera, and the focuser (see Figure 9.3). In addition, a powerful dew-heater system needs to be employed so the operator does not need to keep leaving the house to de-dew Schmidt-Corrector plates or Newtonian secondary mirrors. If the telescope is housed in a dome, as opposed to a run-off roof or run-off shed system, the dome will need controlling from indoors, too, unless the slit is very wide and will allow an hour or two of patrolling near the meridian. The U.K. supernova discoverer Tom Boles uses two Paramount/Celestron 14 systems in a huge run-off roof building, but he also has a Celestron 14 in a dome for patrolling during windy weather. My own Paramount ME/Celestron 14 system uses cables buried under the 40-m length from my study to the observatory. The telescope is controlled via a straightforward 40-m length of high-quality twisted pair (category 5) screened data cable using the RS 232 serial communication port on my Paramount. This works at the relatively slow data rate of 9,600 baud (bits/second)

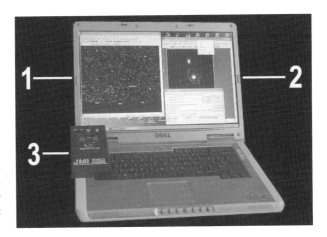

Figure 9.3. The telescope mount, CCD camera, and focuser have to be remotely controlled for a realistic supernova patrol.

because instructions to control a telescope only involve small amounts of data. Various people told me that RS-232 would be "flakey" over that sort of distance but this has not proved to be the case. The motorized JMI moto-focus unit is controlled from indoors, too. In this case, I simply extended the normal hand controller cable from a couple of meters in length to 40 m. Again, I was told this would be very unreliable, but it works flawlessly. Finally, we come to the CCD camera control via the camera's USB port. Here we are dealing with large amounts of data even for the 512 × 512 pixel images from my ST9XE. The USB 1.1 data rate for items other than keyboards and mice is 12 megabits/second. USB 2.0 can operate at up to a blazing 480 megabits/second! Even at the slower rate of my ST9XE, this is more than a thousand times the speed of my serial connection. At these high data rates, the inductance and capacitance of the cable significantly degrades the waveform, and data can be corrupted so badly that zeroes and ones are no longer distinguishable. Therefore, for USB rates an alternative system is needed when the signals are being transmitted more than 5 or 10 m. The solution I adopted was to purchase a device called a USB extender, made by a company called Icron (www.icron.com) in Canada. This device has two component parts called LEX and REX that sit at the PC end and the telescope end (respectively) of my system. The two units work together to eliminate errors between transmission and reception, thus extending the range of my USB 1.1 system to cope with the 40-m range over which I am working. In fact, the Icron Ranger I use is advertised as being suitable for extending USB to at least 100 m. My Serial/Long cable/Extended USB system has worked flawlessly (well, almost!) with my ParamountME/JMI focuser/ST9XE for the last 3 years, so I have no intention of changing it. Modern Paramount MEs feature USB control that, if used, would necessitate a second USB extender or one that could reliably cope with controlling two devices simultaneously. Some modern PCs with no serial ports need a USB port replicator to talk to original serial port Paramounts like mine. I have heard that the USB/serial port replicators made by I/O Gear are the most reliable by other USB Paramount users.

However, different amateurs use completely different automated systems. Supernova patroller Ron Arbour has a very neat observatory in which a warm control room is built onto his LX200 observatory. Ron uses customized software written by himself and a friend and, as the telescope sits only a couple of meters from the

PC, no long-distance cabling is required. Mark Armstrong's system is virtually the same as mine (i.e., it uses an Icron USB extender for the camera control). However, Tom Boles uses a local area network and pcAnywhere from Symantec to control his three telescopes. Windows XP Pro has a remote PC local area network capability built in, but if the PCs are a long way away, suitable hardware is needed to cope with the distances involved. Of course, it is possible to leave an observatory PC and a telescope chugging away while you are asleep in bed. Indeed, initially remote telescope control was cited as being for this very purpose. However, in a climate like the United Kingdom, where you can easily have a completely clear sky 1 minute and rain the next, and where dew can beat even the most powerful dew heaters, most amateurs prefer to oversee proceedings. They are then able to dash out, at a minute's notice, if they suspect something is not quite right. Few U.K. amateurs that I know can sleep easy in their beds with a telescope chugging away unattended. Okay if you live in Arizona maybe, but. . . . Also, spotting a supernova as the image downloads is often critical to clinching the discovery before your closest rival. I know of one supernova patroller who sleeps on a mattress on the floor of his downstairs control room when patrolling so he can "cat-nap" and quickly run outside if it starts to rain. It is perhaps worth mentioning here that the only disadvantage with quality German equatorial mountings like the Paramount ME is that they cannot track far past the meridian without the risk of telescope tube hitting the pillar. This problem is impossible to avoid when the telescope is pointing close to zenith stars (i.e., stars at a similar declination to your latitude), and modern mountings will actually prevent you from tracking more than about 20 minutes of time beyond the meridian. Thus, if you try tracking or slewing through the meridian such that the tube is becoming lower than the counterweights, your telescope drive will either cut out or, as with the Paramount ME, the telescope will normalize itself. By normalize I mean that the software will rotate the declination and right ascension axes by 180° to point the telescope at the same target but to avoid hitting the plinth. With the ME it does this automatically while safely ensuring that the telescope never points below the horizon. Clever, eh? The only problem with this is that with such a drastic rotation about each axis, it is very easy for wires and cables to become snagged and, potentially, damage to occur. The ME will stop slewing and cut out when the current drawn becomes too high, but it is a powerful mount and some damage could occur while you are asleep in your control room. So, with automated patrols it is a good idea to plan the patrol in advance so you are awake when a normalizing maneuver occurs! Another aspect to this is that if you are patrolling close to the meridian, you do not want to switch sides more than once a night as it simply wastes time as you wait for a minute or so for the change to take place. Plan your patrols carefully to avoid loads of east–west slews near the meridian, unless you have a fork-mounted telescope. Unfortunately, a Paramount fork was not available at the time of writing, mainly because a fork mount can only be built for one specific aperture and the long tines of a fork introduce undesirable mechanical flexure. Remember, too, that even a mount as good as the Paramount may show a few arc-minutes of pointing error after such a drastic repositioning, especially if you have a Schmidt-Cassegrain that is very prone to mirror-flop. It is also worth mentioning at this point that supernova patrollers do *not* spend all night slewing huge distances around the sky. This is a time-consuming event and unnecessary. When you have thousands of galaxies in your database, then unless you are patrolling a galaxy close to the plane of the Milky Way there will always be others a few degrees away. Even with a Paramount you can only slew at 5 degrees per second, so huge slews will eat up valuable patrol time and, over years of patrolling tens of thousands of galaxies, long slews will cause unnecessary wear on the worm and wheel.

Software for the Task

So, you have a computerized GO TO mounting and a CCD camera and want to discover supernovae. What software is available to help you to control the telescope and to check the images? The most popular telescope control solution is that provided by Software Bisque; the same team developed the Paramount mountings to do full justice to their astronomy software. They offer what is arguably the best planetarium package for the serious amateur astronomer, namely, *The Sky*.

The Sky has always been the planetarium package that advanced amateur astronomers have regarded as "the gold standard." It has never been an inexpensive package, although you can now buy it in cut down "Serious Astronomer" and "Student" editions. *The Sky* is probably the only package here that can justify the term *professional* as it is used in thousands of advanced amateur and professional observatories to control telescopes that are involved in real science. This is a planetarium package that leaves nothing scientific out: new discoveries are seamlessly uploaded from the Web. All the scientific data you could ever want is here, if your aim is to explore the night sky. Unlike some competing packages, you will not get dozens of animated tours and you will not feel like you are in a spaceship. The current version 6 is much prettier than version 5 and *The Sky 6* display combines beauty and science perfectly. This is the package that the serious amateur astronomer will want to control his or her telescope with total confidence. It may not be quite as intuitive to use as, say, a package like Project Pluto's *Guide 8.0*, and it is a lot more expensive, but for such a powerful piece of software it is quite easy to master. Where *The Sky 6* really excels for supernova patrolling is when it is combined with Software Bisque's other packages *CCDSoft* (a CCD camera/image processing program) and *Orchestrate* (a scripting program for automated supernova patrols, etc). All three packages running on one PC are shown in in Figure 9.4. A fourth package, *T-Point*, provides sophisticated tracking and pointing refinement software for your telescope so that the mount can learn to correct pointing errors.

Efficiently integrating a planetarium package, which controls your telescope, with a CCD camera package, which takes the images, is essential for automated supernova patrolling. If exposures of 30 to 60 seconds are enough to record faint supernovae, the last thing you want to be doing is switching between planetarium and CCD windows on your PC and spending 30 seconds per image just battling with the mouse-clicking activity! Supernova patrolling is carried out at night, when most people are flagging anyway and you want to have the whole process automated. Battling with a PC, moving to different galaxies, and then switching to control the CCD software can be rather tedious after a dozen galaxy images have been taken, never mind a thousand! Obviously some kind of list of commands is required to instruct the telescope to go to another galaxy and then instruct the camera to take an exposure of a certain length and continue this activity for hundreds of galaxies. In the case of *The Sky* software suite, this is carried out by the *Orchestrate* Scripting Package (see Figure 9.5), which, as its name suggests,

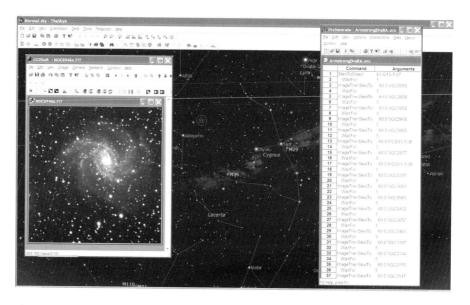

Figure 9.4. *CCDSoft* and *The Sky*, integrated with an *Orchestrate* script file, can run a very efficient supernova patrol from one PC.

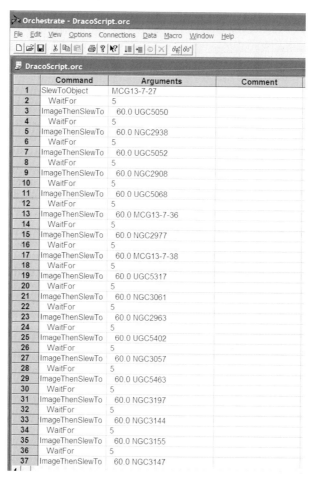

	Command	Arguments	Comment
1	SlewToObject	MCG13-7-27	
2	WaitFor	5	
3	ImageThenSlewTo	60.0 UGC5050	
4	WaitFor	5	
5	ImageThenSlewTo	60.0 NGC2938	
6	WaitFor	5	
7	ImageThenSlewTo	60.0 UGC5052	
8	WaitFor	5	
9	ImageThenSlewTo	60.0 NGC2908	
10	WaitFor	5	
11	ImageThenSlewTo	60.0 UGC5068	
12	WaitFor	5	
13	ImageThenSlewTo	60.0 MCG13-7-36	
14	WaitFor	5	
15	ImageThenSlewTo	60.0 NGC2977	
16	WaitFor	5	
17	ImageThenSlewTo	60.0 MCG13-7-38	
18	WaitFor	5	
19	ImageThenSlewTo	60.0 UGC5317	
20	WaitFor	5	
21	ImageThenSlewTo	60.0 NGC3061	
22	WaitFor	5	
23	ImageThenSlewTo	60.0 NGC2963	
24	WaitFor	5	
25	ImageThenSlewTo	60.0 UGC5402	
26	WaitFor	5	
27	ImageThenSlewTo	60.0 NGC3057	
28	WaitFor	5	
29	ImageThenSlewTo	60.0 UGC5463	
30	WaitFor	5	
31	ImageThenSlewTo	60.0 NGC3197	
32	WaitFor	5	
33	ImageThenSlewTo	60.0 NGC3144	
34	WaitFor	5	
35	ImageThenSlewTo	60.0 NGC3155	
36	WaitFor	5	
37	ImageThenSlewTo	60.0 NGC3147	

Figure 9.5. An *Orchestrate* script file.

orchestrates the operation of the two packages, *CCDSoft* and *The Sky* such that when telescope slewing activity stops, an exposure is started, and when the exposure stops, the next slew commences. All this activity takes place without any intervention from the astronomer, who can just sit back and watch the images downloading. In a cluster of bright galaxies, the images can be quite mesmerizing as they download and are automatically saved.

For those observers familiar with *The Sky* package, enabling *Orchestrate* simply involves the use of four software packages that already come supplied when you purchase Software Bisque's Paramount, namely, *The Sky*, *CCDSoft*, *Observatory*, and *Orchestrate*. The usual procedure is simply to start *The Sky*, power up/initialize the telescope mount, and establish contact between your PC and the telescope mount. *CCDSoft* should then be used to establish a communication link with the CCD camera. At this point, I usually run the small *Observatory* package and simply select "File-New". Then I run *Orchestrate*, go to "Connections" and verify connections to *The Sky*, the telescope, the camera, and the filter wheel that were already specified when *The Sky* and *CCD Soft* were initially set up. A "script file" you previously prepared can then be started once you are 100% happy that everything is working. It is advisable to make *CCDSoft* the active window at these times (i.e., click on the *CCDSoft* window to bring it to the front while *Orchestrate* is running). If all is well, both *CCDSoft* and *The Sky* produce little "server" windows that listen out for the *Orchestrate* controls.

Of course, anyone familiar with software will be well aware that things are often *not* that simple, and different PCs, different software versions, and driver problems can leave you banging your head against a brick wall! The best solution in all these matters is to join a user group (mostly found on *Yahoo!*) to help with these issues. Two heads are better than one and two hundred are far better, especially if the hardware/software designers are on the user group. In this instance, the two most relevant user groups are http://groups.yahoo.com/group/paramount/ and http://groups.yahoo.com/group/SBIG/, both of which have regular discussions of technical hardware and software queries affecting the Paramount or SBIG CCD cameras. But these are just two examples, and it is obviously best to join the user groups relevant to your particular hardware. One problem that is occasionally reported with the *CCDSoft–The Sky–Orchestrate* trio is that if a problem locks the camera up (e.g., it may become very damp after months in the observatory, or the USB extender might have become very damp), *CCDSoft* can completely freeze and cannot be shut down, even with the Windows Task Manager. This sometimes happens when the camera cooling is instigated. In such situations, it is best to bring everything indoors and dry it out, replace the camera's silica gel dessicant, check all the connectors, and then try again the next night. If severe incompatibility problems exist with an SBIG camera, try running SBIG's *CCDOPS* package alongside *The Sky*, but without *Orchestrate*. This will make a fully automated system impossible, but it will at least make your remote control telescope usable again.

The most typical format for an *Orchestrate* supernova patrol script is as follows:

```
SlewToObject        NGC5033
WaitFor             5
ImageThenSlewTo     60.0 NGC4736
```

```
WaitFor              5
ImageThenSlewTo      60.0 NGC5055
WaitFor              5
ImageThenSlewTo      60.0 NGC5194
```

Essentially, this uses commands that *The Sky* translates into slew commands for the Paramount mounting (or any other compliant mounting) and galaxy designations understood by *The Sky*, which it converts into coordinate positions used by *The Sky*. In addition, commands to the CCD camera and their exposure duration in seconds are used, too. So, the example above simply says that the telescope must first slew to the galaxy NGC 5033 and wait 5 seconds for the tracking to settle down after the slew. Then a 60-second exposure is taken before slewing to the next galaxy (e.g., NGC 4736), and so on. While this is happening, the patroller simply watches as *CCDSoft* (set to autosave) saves the galaxy images to the hard disk.

Of course, few amateur astronomers can afford a Paramount mounting. At the time of writing, the ME cost $12,500. However, it is ultrareliable and will track for at least 2 minutes with no perceptible tracking error. However, the software combination of *The Sky*, *CCDSoft*, and *Orchestrate* will work on any modern mounting, like an LX200 or an RCX400, for example. The difference is that, after a million slews, the Paramount will still be working and slewing to arc-minute precision, whereas lesser GO TO mounts will be on their 20th set of gearboxes and will not put the galaxy in the center of the chip.

Other software packages can be used to control telescopes, too. Project Pluto's *Guide 8.0* can do it, as can *SkyMap Pro* and the higher specification and beautiful *Starry Night* programs. The difference with *The Sky* is that it specifically caters for robotic operation and unattended patrolling as it integrates the telescope and camera control with the *Orchestrate* target script.

It has often been said that patrolling for supernovae is not difficult, and that is very true. In essence, galaxies are being imaged using a modern telescope's GO TO facility and the supersensitivity of modern CCD cameras. The advent of microprocessor telescope control, planetarium software, and affordable CCD cameras has placed the backyard amateur in a more powerful position than any professional observatory prior to the 1980s. The research of thousands of electronics engineers and scientists and the power of mass production has meant we don't have to do the tricky part (i.e., building the equipment): it has all been done for us. However, while taking hundreds of galaxy images is not difficult, checking them still is. This is the part of the process that separates the men from the boys, or, rather, the obsessed patroller from the normal, sane individual. For most people, the thought of searching through 5,000 images manually, in a few days, looking for a new speck of light, is horrendous. It all depends how badly you crave that success and that fame. As with any aspect of life, if you want to achieve something extraordinary, you have to be prepared to endure more misery than the vast proportion of other people. You literally have to be obsessed with the idea of astronomical discovery. Unlike in many other areas of life where a huge obsessive effort is involved (e.g., outstanding athletes, tennis players, and footballers), there is no financial reward at all. *You* spend thousands of dollars on your telescope and your CCD camera and you almost kill yourself checking the images. But all you get as a reward is one discovery, which the vast proportion of people in the world will never know or care about. So, by normal standards, being barking mad helps in supernova patrolling! If a flawless system of checking images was available, then more people would have a go at this unusual sport. But then everyone would be discovering supernovae and it would not be so prestigious amongst the amateur community.

So, if the nightmare part is checking the pictures, what software techniques do amateurs use to inspect their galaxy images?

Blink Comparison

The traditional way of checking for new astronomical objects is by blink comparison. This goes back right to the days of checking glass plates (e.g., as in Clyde Tombaugh's famous hunt for Pluto). The human eye and brain is exceptionally good at spotting rapid change; something new blinking on and off. Presumably this skill has developed over countless millions of years as the trick for avoiding predators creeping up on you (or spotting a meal coming your way!). The detection of a sudden movement is essential for survival. So, if you can get an old (master) image of a galaxy and a new one and blink the two, a new supernova will stick out like a sore thumb. This technique works very well. However, think for a moment about what we are trying to achieve. The old and new galaxy images have to be aligned perfectly for the technique to work. Otherwise, the entire field will be blinking on and off, as misaligned stars coincide with black regions of sky in the alternate image. So, every galaxy image must be aligned to within a few arc-seconds to make a blink comparison work, especially when the field is full of stars. Also, the successful patrollers face blinking hundreds or even *thousands* of images with corresponding master frames the day after (or during!) a run of clear nights. Any system needs to be slick and intuitive with as few mouse clicks and keyboard operations as possible when hundreds of galaxies are involved.

So are there any existing commercial packages that come close to doing what we require (i.e., simply blinking the images with the minimum of fuss)? Unfortunately, the answer to this is no. There is nothing that works well enough to make blink comparison a formality. Most supernova patrollers I know simply open up a whole batch of patrol images in *CCDSoft* (by setting it as the default package for opening the camera's FITS files) and then load each deep master image one at a time. An additional problem here is that supernovae can lurk deep in the brightest regions of a galactic bulge, so a low-contrast, low-brightness setting for both patrol and master image is often needed to reveal supernovae near the galaxy center. While *CCDSoft* does have a blink comparator function, and it will auto-align two images in a single folder, this operation can take a minute or two to set up. This is not a problem if you are blinking one image, but just too much hassle if you have hundreds. Most patrollers I know find it quicker to just look at the master and patrol image provided there are not that many stars in the region of the galaxy. Richard Berry and James Burnell's *AIP4Win* software has an alignment and blink comparator function but, again, it is not simple enough to check hundreds of images with minimal user intervention. Essentially, the ideal supernova checking software would do the following:

1. Automatically load the most recent galaxy images (ideally, as they are exposed).
2. Automatically find the master reference frame for that galaxy (by file name).

3. Align the master and reference frames to pixel accuracy.
4. Blink the frames at a high-contrast setting for both.
5. Blink the frames at a low-contrast setting for both (galaxy core).
6. After, say, 10 seconds, move on to the next galaxy image unless paused by the observer.

In the above list, I am largely assuming that the observer does the checking visually while sitting at the PC. However, the ideal system would work through the frames rapidly/as they were taken and simply alert the observer to potential supernovae. The observer could then study the few frames on which a new object appeared and use his or her intelligence to deduce whether it was a real supernova. Even after an extra bright dot is found, there is no guarantee that a supernova has been discovered. There are still plenty of pitfalls for the unwary. While on the subject of supernova checking software, I would like to mention another method that I originally used for comparing images. The Microsoft PowerPoint package is used widely for presenting images at meetings and in lectures. However, its basic function (i.e., to store and rapidly view images), can be used for supernova checking. Images in BMP, TIFF, or JPEG format can easily be loaded into PowerPoint and a text heading can be typed onto each slide page (e.g., "NGC6946"). Your deepest master galaxy images can all be loaded into one or several PowerPoint files with an accompanying text label and the galaxy master image can rapidly be located using the PowerPoint text search facility. Once the galaxy master you want has been acquired, a spare "new slide" space can be temporarily created in the next slot and the new galaxy image can be inserted for comparison. Rotating the new slide to match the master orientation can also be accomplished easily. Hitting the "Page Up" and "Page Down" keys then rapidly alternates between master and image. It is not a perfect blink comparator by any means, *but* it can provide a simple way of checking for obvious supernovae.

So far I have not mentioned freeware for supernova patrolling, but there is one promising package that is being improved even as I type these words. I would like to digress for a few sentences at this point. In 2004, I wrote an article for *Sky & Telescope* that appeared in that year's October issue. In it, I described the phenomenal achievements of the U.K.'s three multiple discovery patrollers, namely Mark Armstrong, Tom Boles, and Ron Arbour. Shortly afterwards, I was contacted by a number of people who thought they might be able to write software to check galaxy images for supernovae automatically. The third person was already known to me, namely Dominic Ford of Cambridge University. Dominic was the meetings recorder at British Astronomical Association meetings, and we briefly discussed the requirements for an efficient supernova checking program shortly after my *Sky & Telescope* article. Subsequently, Dominic contacted Tom Boles to see what was really required. Anyway, to cut a long story short, Dominic has produced an impressive software package called *Grepnova* that allows master and patrol images to easily be loaded, and it also automatically aligns the images prior to blinking them (see Figure 9.6). The software is still being refined as I write these words, but it can currently be accessed at Dominic's site at http://www-jcsu.jesus.cam.ac.uk/~dcf21/astronomy.html. Alternatively, a Google search for "Grepnova" or "Dominic Ford" will pick up this utility. It is a very useful package and is still being improved.

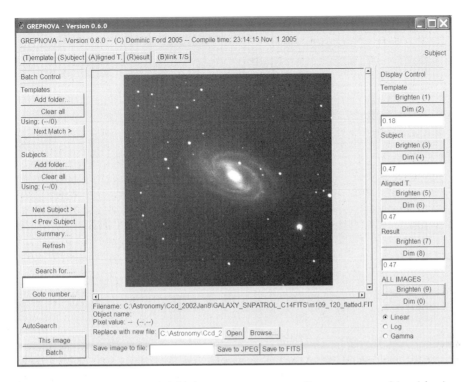

Figure 9.6. Dominic Ford's slick blink comparator program *Grepnova* is a useful tool for the supernova patroller.

Avoiding False Alarms

So, if a new star-like object appears on your galaxy image, what can it be, apart from a new supernova?

Hot pixels and cosmic ray hits are the first trap for the beginner. These can cause bright markings on the CCD image, but the experienced imager will spot them right away. Stars have bright cores that gradually fade into the background sky as you move away from the center. Hot pixels typically occupy one single pixel and look impossibly sharp at long image scales of 1 to 2 arc-seconds per pixel. Cosmic ray hits (i.e., energetic particles that have made it through the earth's atmosphere to hit your detector) also look most unstellar. In all cases where a supernova is suspected, confirmatory images on two separate nights are required anyway, which will weed out these types of false alarms. Also in this category are asteroids, which can cross the field and look remarkably like supernovae. These are especially troublesome in the ecliptic constellations, and as the galaxy-rich constellation of Virgo is crossed by the ecliptic, considerable caution should be exercised when patrolling this region. Again, a confirmatory image on a second night will show the suspect has moved. The Minor Planets Center feature an online asteroid checker (Figure 9.7) for weeding out asteroids from your images. This is located at http://scully.harvard.edu/~cgi/CheckSN but a planetarium package like *The Sky*, or *Guide*, when set up correctly, will also show any asteroids crossing the field.

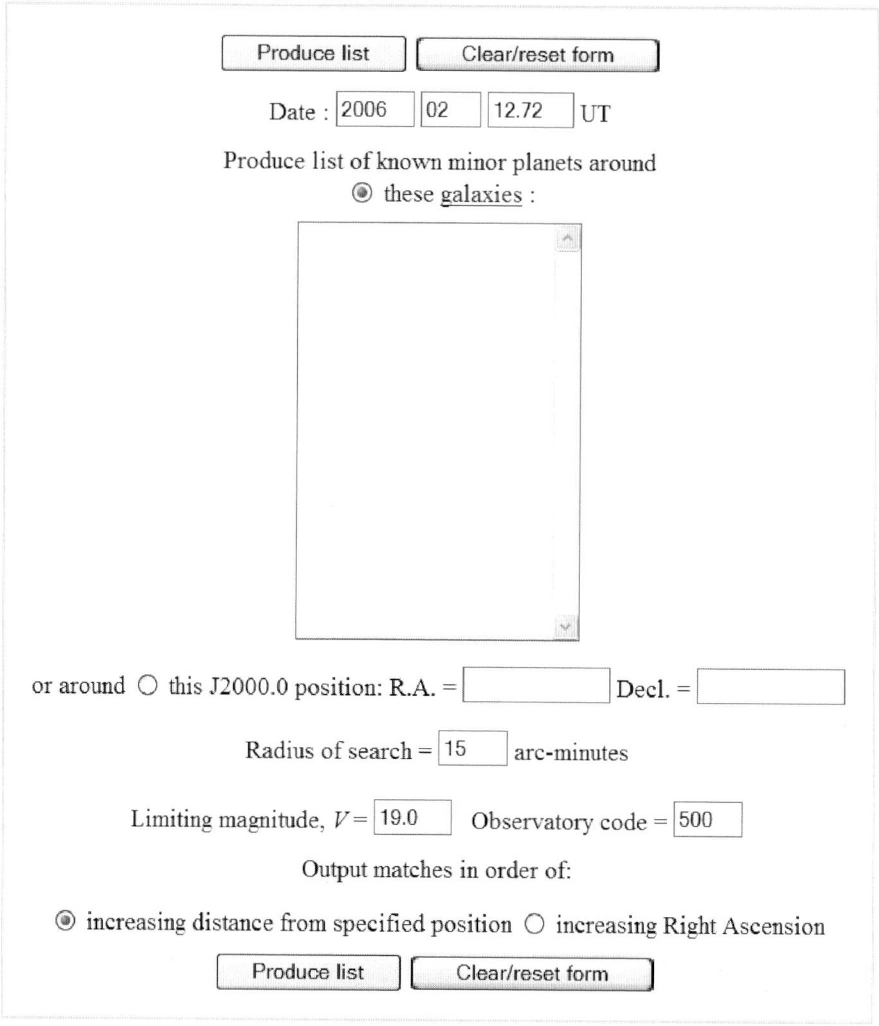

| Produce list | Clear/reset form |

Date : 2006 02 12.72 UT

Produce list of known minor planets around
⦿ these galaxies :

or around ○ this J2000.0 position: R.A. = [   ] Decl. = [   ]

Radius of search = 15 arc-minutes

Limiting magnitude, V = 19.0 Observatory code = 500

Output matches in order of:

⦿ increasing distance from specified position ○ increasing Right Ascension

| Produce list | Clear/reset form |

Figure 9.7. The Minor Planets Center SN Candidate Checker Web page can weed out annoying asteroidal vermin that might look like supernovae!

Other potential sources of false alarms are stars within our own Milky Way, between us and the galaxy, that are variable. Admittedly, such variables are rare, but not unheard of. Of course, you may have discovered a new variable star, in which case you have still made a discovery and the spectra will reveal what you have actually found. One example of a famous Cataclysmic Variable that is very close to the field of a galaxy is AL Com, which is in the same CCD field as M 88 in Coma Berenices. CP Draco is another case. It is close to the galaxy NGC 3147 and has been mistaken for a supernova in the past. Variable stars are especially dangerous as they will not move from night to night, as an asteroid would. The SIMBAD database is an invaluable astronomical resource in this regard. It provides basic data, cross-identifications, and bibliography for astronomical objects outside the solar system and can be queried by object name, coordinates, other criteria (filters), and lists of objects. It currenly contains data on 3,647,505 objects. There are various mirror Web sites for SIMBAD but the main site is at http://simbad. u-strasbg.fr/Simbad.

The other trap that the beginner can so easily fall into is rediscovering a supernova that someone else has discovered a few days or weeks earlier. Remember, several hundred supernovae are discovered each year, and while many are extremely faint, there are often a dozen or more 17th mag supernovae that are still within amateur patrol range even a month or two after discovery. They do not move from night to night and they do not occupy a single pixel. They are supernovae, but they have already been bagged. A quick check of the CBAT recent supernovae pages at http://cfa-www.harvard.edu/iau/lists/RecentSupernovae.html will keep you up do date. Alternatively, Dave Bishop's site contains data and images of recent discoveries at http://www.rochesterastronomy.org/supernova.html.

An invaluable aid for the beginner is simply having the e-mail addresses of any successful supernova patrollers in your country. In my experience, such people are only too keen to help out and they have almost always secured high-quality recent images of any bright galaxy you may have a query about. They are also invariably keen to capture a confirmatory image on the next night if you are clouded out. Incidentally, even in such a small country as the United Kingdom with only three dedicated patrollers (Mark Armstrong, Tom Boles, and Ron Arbour), there have been quite a few joint discoveries where two U.K. patrollers imaged the same galaxy within hours of each other. Out of Tom's 103, Mark's 73, and Ron's 16 discoveries, Mark and Tom have shared four discoveries and Mark and Ron were attributed joint discovery status for two more. Apart from these three patrollers, only two others were successful in the early years of the U.K. successes, namely Stephen Laurie in April 1997 and Steve Foulkes in January 2000.

Master Images

I thought I would just say a few words about obtaining master reference images as the lack of a good master image has, historically, caused embarrassment on many occasions. The first thing I would like to say is that if I had a dollar for every time a nova, comet, or supernova discovery has been claimed on the first night of use with a brand new camera, or telescope, I would be a very rich man. You have to get familiar with your equipment before you try discovering anything. Every CCD detector has a slightly different sensitivity, and large variations in quantum efficiency exist across the spectrum from the far violet to the infrared. Use a new camera, and on that first night new stars will appear where there was nothing with the old camera. Use your first ever CCD camera, and the stars you see will look dramatically different to those from the professionals and from other amateurs. When you start serious patrolling, you must check your patrol images against masters *taken with the same telescope and the same camera*. In addition, your master exposures should be slightly longer than your typical patrol shots so that on nights of really excellent transparency when seeing is good, star images are tight and focus is perfect, you still have a master that is as deep as your sharpest patrol image. Patrol images taken under excellent conditions, when everything goes right, can reveal myriads of faint stars that a normal patrol image does not show (i.e., dozens of potential supernova suspects). You have been warned! If you find that one of your patrol images does go deeper than a master reference image, then that patrol image should become the new master image. Also, never, ever delete images of galaxies. When capturing thousands of galaxy images, it may be tempting to erase them after a while to free up hard disk space. It is far more sensible to simply archive old images onto a CD. You just never know when you might want to refer back to a previous image, as something in the back of your mind tells you there was a suspect in this particular galaxy a year or so ago.

Astrometry

Astrometry is the science of measuring the precise position of an astronomical object with reference to the surrounding stars. An astrometric measurement is essential when you report your discovery to CBAT (Central Bureau for Astronomical Telegrams). Years ago, astrometric measurement was a tedious process involving a measuring engine that precisely measured the positions of stars on glass plates by using x and y motions controlled by micrometer barrels as the measurer viewed the plate through a microscope fitted with crosshairs. Thankfully those days are now gone, and it is a formality to measure a new object's position on a CCD image with software and a digital star catalogue. The aforementioned Software Bisque *CCDSoft* program will carry out accurate astrometry. The integration between *The Sky* and *CCDSoft* enables automatic field recognition to work (usually!). In other words, if the user ensures that the image has north at the top *and* the image scale in arc-seconds per pixel is correct *and* the user supplies the correct right ascension/declination information, there is a good chance that the stars in the database of *The Sky* can be matched to the stars in the FITS format image. However, if the image is upside down or the image scale is wrong or the field is particularly barren, a successful match may not occur. The sequence of events is that the user loads the FITS format image into *CCDSoft* with north at the top and with *The Sky* software also running. On the *CCDSoft* menu, the "Research – Insert WCS/Auto Astrometry" option is selected and, assuming the object's equatorial coordinates shown are correct (they are derived from the FITS header, but can be overruled) and the image scale is correct, the "Continue" option is selected. If all goes well at this point and stars are auto-identified, the "View-Cursor Information" will then show the vital data and, using the astrometric icons such as "Mark Centroid" will show you the RA and Dec positions of specific stars. *CCDSoft* also has photometric options for estimating the new supernova's magnitude. If you have extra databases, such as the U.S. Naval Observatory USNO A2.0 CD-ROMs or the UCAC 2.0 on your hard disk, and enable the faint stars option, more stars become available.

Although *CCDSoft* works well with *The Sky*, my personal favorite astrometry software is *Astrometrica* by Herbert Raab (see Figure 9.8). *Astrometrica* has already been mentioned in the photometry section and is available from http://www.astrometrica.at/. It can be used for a trial period before a nominal fee is sent to Herbert Raab in Austria. Make no mistake, *Astrometrica* is the very best astrometric software available. Not only is it powerful, it is intuitive and virtually free, a superb combination! To measure an image in *Astrometrica*, all you have to do is load the FITS format image with north at the top, go to "Astrometry/Data Reduction" and click "OK". The software will then use the FITS RA and Dec information to

Figure 9.8. Herbert Raab's easy-to-use astrometry software *Astrometrica* provides an easy solution to measuring a new supernova's position.

compare the stars in the image with whatever stars are in the catalogue you have specified for it to look in. I would strongly recommend acquiring the UCAC 2 star catalogues and copying them to your hard disk. At the time of writing, the final fourth UCAC2 star catalogue is about to be issued, but all stars up to +40 Dec are covered in the first three catalogues. You will need to point *Astrometrica* to whatever catalogue you wish to use and tell it where the catalogue lives. The USNO B1.0 data is available on the Web at http://vizier.u-strasbg.fr/viz-bin/VizieR/, and, prior to 2006 has been used to supplement the +40 to +90 Dec data still not covered by UCAC 2. As we saw earlier, *Astrometrica* also gives useful photometric data that can be used to give a reasonable estimate of the brightness of your supernova suspect.

Since 2004, the Lick Observatory has asked that any new supernova discoveries should include offset positions from the nearest bright star. It was always obligatory that offsets from the galaxy center were employed, but now offsets from a bright star are also preferred so that professionals aiming narrow-field spectrographs at the new supernova can have a point source reference nearby. Most galaxy nuclei, especially for big galaxies, are too large and fuzzy to be used for a precise reference. Indeed, even professional catalogues differ by several arc-seconds regarding exactly where their centers are.

Maybe a word or two here about measuring offsets in arc-seconds would not be a bad idea. Let us take the case of the supernova 2005cs in M 51, discovered in 2005 by Wolfgang Kloehr. A precise position for this supernova was measured (on his own prediscovery image) by Peter Birtwhistle as right ascension 13 h 29 m 52.81 s and declination +47° 10′ 35.3″. According to my copy of *The Sky*, the center of M 51 is at 13 h 29 m 53.3 s and +47° 11′ 48″. So what are the offsets of the supernova from the galaxy center? Well, the offset in declination is easy to work out as it is simply +47° 10′ 35.3″ minus +47° 11′ 48″, that is, 10′ 35.3″ minus 11′ 48″ or −72.7 arc-seconds. Working to sub arc-second accuracy for offsets from a fuzzy core are meaningless, so we can say the supernova is 73 arc-seconds south of M 51's core. But what about the east–west offset in right ascension? Here things are slightly more complicated because the actual angular distance subtended by a second of right ascension is smaller as you climb to higher declinations. In fact, you have to multiply by the cosine of the declination. So, for this specific example it is 13 h 29 m 52.81 s minus 13 h 29 m 53.3, or 52.81 s − 53.3 s = −0.49 seconds of RA or −0.49 seconds west. At zero degrees declination, this would correspond with 15 × 0.49 or 7.35 arc-seconds. However, because M 51 is at the high declination of 47.2 degrees, this result needs to be multiplied by the cosine of 47.2 (i.e., 0.68). So the offset is actually 7.35 × 0.68 or 5 arc-seconds west. Actually, for this specific supernova, a number of different offsets were quoted!

Submitting a Discovery Claim

Posting a discovery claim to the Central Bureau for Astronomical Telegrams (CBAT) in Boston, Massachusetts, the world clearinghouse for astronomical discoveries, is *not* something to be considered unless you are absolutely 110% sure that you have made a discovery. From a beginner's viewpoint, you really need to convince the top supernova patrollers in your country that you have found something before you even consider mailing CBAT. Experience and reputation is everything in this game. If your claim turns out to be a false alarm, then you will not endear yourself to real discoverers and you cannot expect to be treated without suspicion if you make another claim at some future time. The discoveries of the U.K.'s Tom Boles and Mark Armstrong are impressive enough on their own, but doubly so when you realize that no false alarms were generated by them in their decade of patrolling.

CBAT is keen to help potential discoverers to weed out any spurious discoveries before they arrive at their clearinghouse. Although the facility is strongly associated with the Harvard University, which has an observatory, they do not have the facilities to chase up discoveries. So CBAT has issued extensive guidelines for discoverers on its excellent Web pages. Even before you think you have discovered a supernova, you should check their Web pages at http://cfa-www.harvard.edu/iau/HowToReportDiscovery.html and http://cfa-www.harvard.edu/iau/DiscoveryInfo.html.

Even if you do have a genuine supernova discovery, it needs to be reported in the correct format. The name and position of the host galaxy needs to be reported as well as the magnitude of the discovery and the time of the observation to an accuracy of 0.01 days. A precise astrometric measurement of the supernova must be included in the report, to a precision of 0.01 s in RA and 0.1″ in dec as well as the offset in arc-seconds (east, west, north, and south) from the galaxy's center. While a 100% confident observation from the discovery night can be reported instantly, especially if you are already a successful discoverer, a second-night confirmation is essential before the discovery will be accepted. So a second night report must be submitted. It is also vital to state that you have compared the discovery image to previous images you have obtained (quoting the time and date) and that you have eliminated existing supernovae and all minor planets in the area. A check should also be made in case a faint galaxy has an active nucleus (i.e., whether it is in the blazer/quasar category). A catalogue of quasars and active galactic nuclei can be found at http://www.obs-hp.fr/www/catalogues/veron2_10/veron2_10.html.

Only after all of these checks have been made can a discovery claim be justified. As an example of how one of the world's top supernova discoverers reported his 58th supernova, in which I played a tiny confirming role, I reproduce below the e-mails from Mark Armstrong to CBAT on the nights of April 6/7 and 7/8, 2004.

Dear Dan

PRELIMINARY REPORT

I have a suspected sn on a single image in NGC 3786 (R.A. 11h39m42.55s Decl +31o54m33s). Astrometry: 2004 04 07.11966 R.A. = 11h39m42s.18, Decl. = +31o54′31″.8 Mag 14.5 (LM of frame 18.5) Offsets 4.7″ W and 1.2″ S. There is no trace of the suspect on my previous images from 2000 Dec 30 (LM 19.5), 2003 May 25 (LM 19.5) and 2003 Dec 7 (LM 19.0). Nothing on POSS-11 plates. The CBAT MP checker was negative. The recent supernova page is negative too.

It was difficult to estimate the mag of the suspect due to the proximity to the galaxy core. The software tends to add a bit to the measure from the core. It could be nearer to mag 14. Note sn 1999bu was in this galaxy and also close to the core but the astrometry and offsets differ, thankfully!

I will get a second night.

Best wishes,

Mark

Dear Dan

Martin Mobberley and myself have confirmed the suspect in NGC3786 this evening. It appears to have brightened slightly but difficult to be sure due to its proximity to the core. LM of frame 19.0 Here is the formal report-M. Armstrong, Rolvenden, U.K. reports his discovery of an apparent supernova (mag. about 14.3) on unfiltered CCD images taken on Apr. 07.120 and 07.847 UT with a 0.35-m reflector. The new object is located at R.A. = 11h39m42 s.18, Decl. = +31o54′31″.8, which is approximately 4.7″ west and 1.2″ south of the center of NGC 3786. Nothing was visible at this position on Armstrong's images taken on 2000 Dec. 30, 2003 May 25 and Dec. 07 and on Palomar Sky survey red and blue plates. Martin Mobberley, Cockfield, U.K. confirmed the suspect on an unfiltered CCD image taken on Apr. 07.831 UT.

Best wishes,

Mark

An IAU Circular was issued announcing the discovery a day later:

Extract from IAUC 8736:

SUPERNOVA 2004bd IN NGC 3786

M. Armstrong, Rolvenden, U.K., reports his discovery of an apparent supernova (mag about 14.3) on unfiltered CCD images taken on Apr. 7.120 and

Chapter 10

The Discoverers Themselves

This chapter does not describe all of the world's amateur supernova discoverers, just some of the very best ones. There are plenty of other discoverers who I have not described in detail, but who have also checked through tens of thousands of images to make their discoveries. However, I do feel that the patrollers described below are not only amazingly prolific but incredibly professional, too. I will start with the biggest name of all, namely Bob Evans, the man who searches visually and inspired all those who followed him.

The Rev. Robert Evans: Doing It Visually

First discovery: 1981A in NGC 1532
Number of discoveries: 46
Location: New South Wales, Australia
Equipment: 25, 30, and 40 cm Newtonians

Although Bob Evans, shown in Figure 10.1, was not the first amateur to discover a supernova (he was the fourth) and although there are now four amateur astronomers who have discovered more, namely Armstrong, Boles, Puckett, and Schwartz, Bob Evans is still the only truly legendary figure in supernova discovery. To discover supernovae visually, simply relying on your memory of the galaxy fields, and to haul a telescope manually to each object, with no GO TO facility, is awesome. Not only this, but Evans has discovered some absolutely cracking bright supernovae: a result of patrolling the brightest galaxies. If you are not too keen on patrolling 10,000 galaxies, Evans' strategy of just bagging a couple of bright, visually discoverable, supernovae per year may be one to emulate. Remarkably, Evans has discovered three supernovae in the same galaxy, NGC 1559, in 1984, 1986, and 2005. In addition, in the beautiful galaxy NGC 1365, he has discovered two supernovae, in the years 1983 and 2001, and likewise for NGC 1448, where he bagged supernovae in 1983 and in 2003. In 1986, Evans bagged the only discovered supernova in NGC 5128, alias the strong radio galaxy Centaurus A. For most amateur astronomers, discovering a supernova in a Messier galaxy (there are only 39) is about as prestigious as it gets. Most of these are in the Northern Hemisphere

Figure 10.1. The phenomenal visual supernova discoverer, Bob Evans, with his 30-cm Dobsonian-mounted Newtonian. Image: Michael Schwartz.

though (as Messier was a Frenchman). Despite this, Evans has discovered four supernovae in Messier galaxies, namely 1983N in M 83; 1988A in M 58 (jointly with Ikeya, Pollas, and Horiguchi); 1989B in M 66 (jointly with Manzini), and 2003gd in M 74. Sixteen of Evans' supernova discoveries are in the brightest 100 supernovae, and he has also bagged 15 very bright, highly prestigious, Type Ia supernovae, an awesome reputation! Bob Evans appeared on a video tape made by Rob McNaught that I used to distribute for Guy Hurst's *The Astronomer* magazine in the early 1990s. Since then, Evans has written up nearly all of his discovery accounts for *The Astronomer* magazine. They make fascinating reading.

Bob Evans started serious galaxy observing in the 1960s using his 25-cm Newtonian. In the early 1970s, he considered turning the telescope into a photographic instrument to patrol for supernovae but that did not work out. However, Gus Johnson's discovery of the bright supernova 1979C in M 100 inspired Evans and with help from Queensland amateurs and Tom Cragg at the Anglo-Australian Observatory, he was able to make photographic slides of hundreds of galaxies as a master reference library. Evans' first independent discovery (1980N) was not attributed to him because he spotted it, as a mag 12.5 intruder in NGC 1316 (Fornax A) just a day after the Chilean Observatory's astronomer Wischnjewsky claimed it on December 7, 1980. It was a close call, but inspired Evans further. Only 11 weeks later Evans did bag his first official supernova discovery with the 25-cm telescope; the first supernova of 1981, in NGC 1532. That discovery, 1981A, was made on February 24, 1981. Coincidentally, the supernova of the century would be discovered exactly 6 years later on February 23/24, 1987, that is, SN 1987A in the Large Magellanic Cloud. Bizarrely, less than 3 months after Evans missed 1980N, he made up for the one that got away in NGC 1316. Another bright supernova appeared in that same galaxy, at magnitude 12.7, and became Evans' second discovery. Two bright supernovae occurring in the same galaxy within a few months is almost unheard of! Maybe the Reverend does indeed have inside information on the workings of the cosmos!

So, can anyone copy this Australian observer's visual trick? Well, before we get too excited, I think it is important to remember that there is less competition in the Southern Hemisphere, there are more clear nights than in most other countries, and, most importantly, winter nights in Australia are rarely as cold as in most other patroller's countries. However, there is little doubt in my mind that emulating Evans' visual discovery technique does not require a photographic memory. Every human brain has a remarkable ability to recognize patterns and spot something new. This ability has actually been traced to a specific part of the brain, which, if damaged, can result in the victim being unable even to recognize a picture of himself or herself. Amongst a crowd of thousands of Christmas shoppers, we can easily identify a friend or relative (or an enemy for that matter). Other astronomers have also exploited the power of the human memory to make discoveries. Foremost amongst these was the U.K.'s George Alcock who, in the 1960s, decided to build on his already formidable familiarity with the night sky (from three decades of meteor observing) and commit to memory all of the Milky Way stars he could see through his 80-mm binoculars! Essentially, he memorized, in patterns, some 30,000 stars. As a result he discovered five novae in the years 1967, 1968, 1970, 1976, and 1991. The last of these was discovered from indoors, while observing through double-glazed windows, at the age of 78! As someone who knew George, my explanation of how he achieved this phenomenal memorizing feat is simply that he combined a lifetime of sky watching with a nightly ritual and the already formidable powers of the human brain. If you asked him to draw any region in the Milky Way, to, say, 8th magnitude, from memory, he would find it almost impossible. However, if something new appeared, especially something of 7th magnitude or brighter, he would spot the change in a pattern instantly. Most people could not accurately draw their neighbor's face, but they could spot a change in their appearance. Similarly, they could spot something that was out of place in their home, like an ornament that had moved by a foot or so. It is all about spotting changes in patterns. Obviously, any visual patroller has to have good eyesight and a good memory. However, I would place a love of the night sky, ritualistic behavior, and superhuman patience as the top priorities, not a so-called photographic memory. People with photographic memories, excluding the genuinely rare savants, have simply trained their memories better than the rest of us.

Any visual patrol system has to be sustainable over long periods and, because the observer is outside in the dark, damp, and cold, ergonomics has to be a top priority. A telescope with an uncomfortable observing position, one that is complicated to set up, and weighs half a ton, is not suitable for hauling around the sky from galaxy to galaxy. One of the reasons that Schmidt-Cassegrain telescopes have become such a great success is because they can be stored in a small observatory and the eyepiece position moves very little as you move the telescope around the sky. Seemingly trivial issues like this can become all important when trying to find 14th magnitude specks of light in dozens of galaxies each night.

The vast majority of Bob Evans' supernova finds have been made with relatively modest amateur telescopes. First he used a 25-cm Newtonian, then, from January 1986, at Coonabarabran, a 41-cm equatorially mounted Newtonian (provided by the Commonwealth Scientific and Industrial Research Organisation) and, finally, a 31-cm Dobsonian. But he also used a 1-m telescope belonging to the Australian National University at Siding Spring on just over 100 nights from 1995 to 1997 (three supernovae were discovered visually with that instrument, from around 10,000 galaxy patrols). Evans has always made discoveries from New South Wales but his precise observing location has varied slightly. When using the 1-m telescope, he lived close to the Anglo-Australian Observatory in Coonabarabran, and he was a minister of the local Warrumbungle Parish of the Uniting Church in

Australia. In the late 1990s, he moved to a retirement villa at Hazelbrook in the Blue Mountains but kept his 41-cm Newtonian stored at a friend's farm at Mudgee. Initially, Evans' work was with a 25-cm Newtonian. Then he used the 41 cm with occasional access to the 1-m telescope at Siding Spring. However, his recent discoveries from the Blue Mountains site have all been with a 31-cm Dobsonian (or a Dobson mounting Newtonian as he more correctly calls it). From reading all of his postdiscovery reports in the British magazine *The Astronomer* run by Guy Hurst, it would appear that Evans usually works with a list of around 1,000 galaxies (depending on the season) and he works through them at a rate of one or two galaxies per minute, moving the telescope by hand from galaxy to galaxy, using a straight-through finder but no setting circles. Evans has stated that the most galaxies he has ever star-hopped to in a single night was 570, in 10 hours in March 1984. Normally, his observing sessions are only a few hours long. In the first 5 years of serious patrolling (i.e., 1981 to 1986), he made 14 discoveries from 50,000 galaxy patrols (i.e., roughly one discovery for every 3,600 patrols), which is a similar patrol success rate to the best amateur CCD patrollers, except at a slower speed, and Evans' discoveries were much brighter.

Other successful visual hunters seem to use a similar strategy. In the 1996 December edition of *The Astronomer* magazine, the Italian visual supernova hunter Stefano Pesci reported that in a 4-year search program by himself and his Italian colleague Piero Mazza (using 40- and 51-cm telescopes), 30,000 galaxy observations were made on a database of 600 to 800 galaxies within about 130 million light-years. Two supernovae were discovered in that time and three were missed (one by 20 hours, one by 3 days, and one overlooked). Stefano reported that each galaxy was checked in 2 to 3 minutes but only 20 to 30 nights per year were used for patrolling. If my math is correct, this works out at roughly 150 galaxies per night per observer. Crucially, Stefano stressed that he and his colleague just enjoy observing galaxies: the supernovae are a bonus, not the single aim. By sticking to relatively nearby galaxies, they tried to ensure any supernovae would be brighter than magnitude 15.5.

Michael Schwartz: Thinking Big!

First discovery: 1997cx in NGC 3057
Number of discoveries: 33 individually; 258 LOTOSS team discoveries
Location: Oregon andArizona
Equipment: 35 cm SCTs; 50 and 81 cm Ritchey-Crétiens

Michael Schwartz was the CEO of his own software company, Prime Factors Inc., from 1981 to 1997. When he sold the company for a small fortune in 1997, Michael suddenly had the cash to really indulge in his passion for astronomy. This was around the time that the first Paramount telescope mountings were being manufactured and, as we have already seen, Michael realized that a Paramount, coupled to a Celestron 14, and working at an image scale of 1 arc-second per pixel was a formidable supernova searching tool. But, as we shall see, one single C14 was just the first step in Michael's quest. On July 12, 1997, Michael discovered his first supernova with the Celestron 14/Paramount system: a 15th mag Type II example in the galaxy NGC 3057. Just over 2 years later, a joint supernova search was arranged

Figure 10.2. The massive Tenagra II 0.81-m f/7 Ritchey-Cretien of Michael Schwartz: the largest amateur telescope used for supernova discoveries. Image: Michael Schwartz.

between the professional Lick Observatory Supernova Search and Michael's Tenagra Observatory Supernova Search. The joint venture was called LOTOSS and went on to discover 258 supernovae. To date, 33 supernovae have been discovered or codiscovered by Michael outside the LOTOSS collaboration (which ended in 2003). The original Tenagra I C14 telescope in Oregon was soon dwarfed by a massive 0.81-m (32-inch) f/7 Ritchey-Cretien telescope (Figure 10.2), which Michael housed at a new facility in Arizona in 1999. This telescope has been involved in a range of impressive projects, including asteroid discovery and follow-up work as well as the collaborative work with Lick Observatory between 1999 and 2003. Professional astronomers can (and do) buy observing time on this massive instrument from $175 per hour. In recent years, Schwartz has expanded his Tenagra Observatory Supernova Search (TOSS) and gamma ray burst follow-up work to Norway and Australia. Amateur astronomers Odd Trondal and Paul Luckas operate the Tenagra facilities in those countries, using 0.36-m Celestron 14 instruments. His Tenagra observatories home page is at http://www.tenagraobservatories.com/index.htm.

Incidentally, the Tenagra name comes from a Star Trek episode.

Tim Puckett

First discovery: 1994I in NGC 5194 (Messier 51)
Number of discoveries: 140 individually and within his team
Location: North Georgia
Equipment: 35 cm SCTs; 50 and 60 cm Ritchey-Crétiens

Tim Puckett (Figure 10.3), of Ellijay, Georgia, USA, has his name on more supernova discoveries than anyone including the infamous (and abrasive) Fritz Zwicky who discovered 123 supernovae in the photographic era from 1921 to 1973, a discovery span of an incredible 52 years! Tim's first supernova discovery was back in 1994 when he codiscovered the first ever supernova in the whirlpool galaxy M 51 (NGC 5194), close in to the core of that galaxy. His joint CCD discovery with Jerry Armstrong of Atlanta, Georgia (not to be confused with the U.K.'s Mark

Figure 10.3. The ultraprolific supernova discoverer Tim Puckett and his 0.6-m Ritchey-Cretien based in Georgia, USA. Image: Tim Puckett.

Armstrong) was just preceded in time by those of Wayne Johnson and Douglas Millar of the Orange County Astronomers. Reiki Kushida in Japan and the well-known author Richard Berry of Wisconsin made independent discoveries, too. Since that time, Puckett has worked tirelessly to discover as many supernovae as possible with a variety of collaborators. Some of these collaborators have worked with him at his observatory in the North Georgia Mountains while others have simply collaborated by sharing some of the workload from other sites. Unlike so many other top supernova discoverers, Tim Puckett has a day job, too! Tim's early supernova discoveries were made with yet another Paramount/C14 system as well as a massive homemade 60-cm f/8 Ritchey-Cretien telescope. A further instrument, a 50-cm Ritchey-Cretien, has recently been added. Like Michael Schwartz, Tim Pucket has expanded his Puckett Observatory Supernova Search (POSS) to become an international patrol. In addition to his 60, 50, and 36 cm instruments, Tim recently incorporated another 50-cm instrument based at Ajai Sehgal's observatory in Osoyoos, British Columbia, and hopes to add another instrument in South Africa very shortly. All of Tim's remote telescopes return images to him via the Internet. Thus he can image up to 1,600 galaxies per night, similar to the output of Mark Armstrong and Tom Boles in the United Kingdom, with their three Paramount/C14s each. Like Tom Boles and Mark Armstrong, Tim has now secured hundreds of thousands of images in his patrol (800,000 as of 2005) and so, like them, needs to check thousands of images for each discovery. Over the past 12 years he has worked with some 30 collaborators at his observatory, and now abroad, who have shared in his success. His current international arrangement involves e-mailing images to volunteers to check, while he oversees the remote robotic operation of the telescopes. Tim also images comets, and his home pages are at http://astronomyatlanta.com/pages/12/.

Mark Armstrong

First discovery: 1996bo in NGC 673 (The first British supernova)

Number of discoveries: 73; all working alone

Location: Rolvenden, Kent (UK)

Equipment: Three 35 cm SCTs on Paramount mountings

The youngest of the United Kingdom's multiple discovery supernova patrollers, Mark Armstrong (born in 1958) made all of his discoveries from Rolvenden in Kent. Mark (Figure 10.4) works from home as a consultant to the U.K. magazine *Astronomy Now*. During his peak patrol years, from 1995 to 2004, he lived, slept, and breathed supernovae when skies were clear and often ended up with a backlog of thousands of images to check through. Mark named his first asteroid discovery, made while supernova patrolling, 15967 Clairearmstrong, after his wife, who supported him through the peak patrol years. His other asteroid discovery was numbered 44016. As Led Zeppelin are his all time favorite rock group and their guitarist Jimmy Page is his hero, he named that asteroid 44016 Jimmypage. The last time I asked Mark about that asteroid, he told me he still had not found a way of communicating this fact to Mr. Paige! When I interviewed Mark for an article I was writing in 2004, he said he often had Zeppelin or Dylan on in the background while watching his galaxy images download, but after midnight he would have live American football and baseball on. (Despite being British, he supports the Oakland Raiders and Chicago Cubs.)

Figure 10.4. Mark Armstrong, of Rolvenden, U.K., at his control center, checking the images from three Celestron 14 patrol telescopes. Image: Mark Armstrong.

Like many observers, Mark patrols using the Software Bisque suite of *The Sky*, *Orchestrate*, and *CCDSoft*, with a PC for each of his three patrol telescopes, and he has standardized on SBIG ST9XE cameras for each system. Working at f/11, each CCD chip covers 9 arc-minutes at a scale of 1 arc-second per pixel. He has an *Orchestrate* script for different constellations, each script containing hundreds of galaxies. With more than one Paramount, he does not have to worry too much about normalizing the German equatorial mountings when they track past the zenith: they all look at different constellations! While sitting indoors in the warm might seem like an easy life, the checking workload is mind-boggling. From 1995 to 2004, Mark personally examined more than a third of a million galaxy images and observed on more than 1,100 nights for his 60 discoveries in that period. In 2003, he discovered 16 supernovae but, frustratingly, just missed 13 more that were bagged by the Lick and Tenagra patrols. These near misses reduced his discovery stats to one supernova for every 5,000 images for 2003. That year he observed on 109 clear or partly clear nights and exposed 83,385 galaxy images: that's an average of 765 images per clear night. But Mark's two newer Paramount systems (the MEs) really delivered the goods in that year, bagging one discovery for every 2,900 images. Mark is pictured with his final three Paramount MEs and Celestron 14s, in 2004, in Figure 10.5.

When a run of clear nights occurs patrolling can be gruelling, especially if you have an equally keen rival in the same country with an almost identical discovery tally. To quote Mark precisely from that period: "In September 2003 I hit a wall. I just had to have a break. It had been clear for three months. It was just getting silly. What I desperately needed was reliable, automated, image subtraction software, to subtract the master from the patrol image, allowing for sky conditions, and alert me to something new. It would make my life so much easier. I will continue to patrol until we know all we need to know about supernovae. I would like to see the major patrollers pooling resources and have a co-ordinated patrol. My dream is to move to the USA, either New Mexico or Arizona, to get the clearer skies and to go to the ball game!" Mark currently has 73 supernova discoveries to his credit. One of his brighter finds is shown in Figure 10.6.

Mark and our next subject, Tom Boles, are pictured together in Figure 10.7.

Figure 10.5. Mark Armstrong with his three Celestron 14/Paramount ME patrol systems. Image: Mark Armstrong.

Figure 10.6. One of Mark Armstrong's brightest supernova discoveries, supernova 2004bd in NGC 3786: a 14th magnitude Type Ia discovery. Image: Mark Armstrong.

Figure 10.7. Tom Boles (left) and Mark Armstrong at the BAA exhibition meeting in Cambridge, England, in June 2004. Image by the author.

Tom Boles

First discovery: 1997dn in NGC 3451

Number of discoveries: 103; all working alone

Location: Coddenham, Suffolk. U.K.

Equipment: Three 35 cm SCTs on Paramount mountings

Being born in Glasgow, Tom Boles is, strictly speaking, a Scottish supernova discoverer even if his observatory is in the small Suffolk village of Coddenham (less than 30 miles east of my observatory). From 2003 to 2005, he served a 2-year stint as the 58th president of the British Astronomical Association (BAA). Tom is a retired telecommunications manager but started his working life as a telescope maker with the Glasgow firm Charles Frank. After discovering his first few supernovae while still in full-time employment, Tom decided to relocate from Northamptonshire to Suffolk: a move of only 80 miles, but it meant much darker skies and, initially, a few clearer nights. After a year installing his old LX200 dome and a massive new run-off roof observatory in the farmer's field alongside his house, Tom was back in action. Sir Patrick Moore officially opened the new observatory on August 19, 2001 (see Figure 10.8). At the new site, Tom has a Paramount GT1100 and a GT1100S, both with Celestron 14 optics, installed in the run-off building (Figure 10.9). He also now has a third GT1100/C14 system in his old dome (Figure 10.10). The dome-based system is invaluable for windy nights when the run-off roof telescopes cannot be used. Tom told me: "I decided at Northampton that observing outside all night in winter, in a freezing cold dome was not a sensible plan for a man in his fifties. Everything is done remotely now." During 2003, Tom's total supernova discoveries overhauled Mark's, a feat that had seemed impossible only a couple of years earlier but has, undoubtedly, kept the friendly rivalry going. By expanding his galaxy database to slightly fainter galaxies, Tom had managed to find a niche for discovering mag 17 supernovae that many of the other amateur patrollers missed. This meant that perfect focusing and tracking was essential, as was a return to longer, 60-second exposures; but it paid off and Tom discovered an incredible 30 supernovae in 2003.

Figure 10.8. Tom Boles, with Sir Patrick Moore, at the opening of Tom's new observatory in Coddenham, Suffolk, on August 19, 2001. Image by the author.

Figure 10.9. Tom Boles with two of his Paramount/Celestron 14 systems in a large run-off roof building at Coddenham in Suffolk. Image: Tom Boles.

Figure 10.10. Tom Boles with his oldest Paramount GT1100/Celestron 14 system, inside a dome at Northampton. This system is still used during windy conditions from his Suffolk observatory, when the run-off observatory telescopes are too exposed to wind vibration. Image: Tom Boles.

Unlike Mark's constellation-based patrol, Tom's *Orchestrate* scripts are along lines of right ascension, patrolling downwards in declination, starting at the high-Dec galaxies that the Lick KAIT telescope can't reach (above +70 Dec). This strategy helps to avoid the Paramount's crossing the meridian and normalizing themselves. Like Mark, Tom found that the late summer months of 2003 pushed him to the limit. "From July to September 2003 there were a ridiculous number of clear nights. I was praying for cloud so I could get some sleep and dreading Rita telling me she could see Mars through the window! I decided I really shouldn't patrol on her birthday (September 26) but kept sneaking a look through the curtains and then feeling guilty as it was clear; but I had a 1,000 galaxy backlog anyway!" Tom currently has a massive 103 supernova discoveries to his credit. He also has a Web site at http://www.coddenhamobservatories.org. One of Tom's brighter supernovae, SN 2002bx, is shown in Figure 10.11.

Figure 10.11. One of Tom Boles' supernova discoveries, imaged by the author: Supernova 2002bx in IC 2461.

Ron Arbour: 16 Bright Supernovae and Counting

First discovery: 1998an in UGC 3683
Number of discoveries: 16; all working alone
Location: South Wonston, Hampshire, U.K.
Equipment: 30 cm SCT

Unlike Mark and Tom, Ron Arbour has been a big name on the U.K. astronomy scene since the 1970s. His dedication to amateur astronomy has been total over the past 30 years and he was responsible for the formation of the British Astronomical Association's (BAA) Deep Sky Section, the Astrophotography Section, and the Campaign for Dark Skies. Ron and his wife, Pat, have moved house three times just to escape from increasing light pollution. They now live at South Wonston in Hampshire. Ron is a rare specimen in 2006: a telescope builder and a mirror-maker who was a force to be reckoned with even when film and cold cameras were the state-of-the-art. In the 1980s, Ron built an incredible 16-inch Newtonian that could track perfectly for 5 minutes unguided, 15 years before the Paramount arrived on the scene. The 16-inch used a friction roller drive and a friction roller gearbox and, from the mid-1980s, patrolled for supernovae using film. The telescope was the sole subject of Patrick Moore's *Sky at Night* program in October 1985. Very reluctantly, Ron switched to a commercial 12-inch LX200 for patrolling in 1997 and still uses the same, highly modified instrument, together with a Starlight Xpress SXV H9 CCD camera for patrolling today. Ron's LX200 drive system has been virtually rebuilt to give better tracking, slewing, and reliability. Ron's telescope and observatory are very similar to the system operated by Berto Monard in South Africa. Unlike Mark and Tom, Ron has the main computer controlling the patrolling in the same building as the telescope. It is a very compact and friendly system that any proficient handyman could construct, as shown in Figures 10.12, 10.13, and 10.14. It is also within the financial grasp of many keen amateur astronomers,

Figure 10.12. Ron Arbour's observatory at South Wonston in Hampshire, England. A control room is conveniently situated adjacent to the observatory. Image by the author.

Figure 10.13. Ron Arbour's compact control room from which his 30-cm LX200 SCT is controlled. Image by the author.

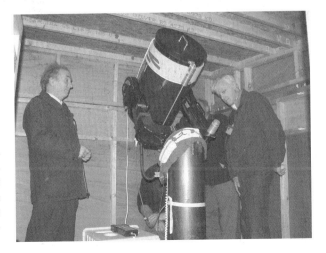

Figure 10.14. Ron Arbour (left) demonstrates his highly modified 30-cm LX200 supernova patrol telescope to some visiting amateurs. Ron Johnson (far right) looks on. Image by the author.

Figure 10.15. One of Ron Arbour's discoveries, imaged by the author. This was the bright supernova 2003ie in the beautiful galaxy NGC 4051.

unlike the three Paramount/Celestron 14 systems of his countrymen Mark and Tom.

For unattended robotic operation of the LX200, Ron uses one piece of home-grown software. "My friend Dave Briggs has written all the image acquisition and display routines and anything that requires assembly language while I have written the BASIC routines for telescope control and astro-navigation."

With a single off-the-shelf instrument and a tenth of the financial outlay of a triple Paramount/C14/ST9XE system, Ron has a different strategy from Mark and Tom. He concentrates on bagging the prestigious bright supernovae and has been very successful with this approach. More successful in fact than his U.K. rivals who mainly capture mag 16 to 18 supernovae. Bagging bright supernovae requires shorter exposure times, so more galaxies can be farmed "and even KAIT can't patrol all my galaxies every night." Twelve of Ron's 16 supernovae have been in bright NGC galaxies. One of these is shown in Figure 10.15. Being retired from his job as a technician at Southampton University, Ron has dedicated all his spare time to patrolling. Recently, Ron purchased the MKS 4000 Paramount electronics from Software Bisque and is using it to control his famous 40-cm friction-drive Newtonian.

Berto Monard: 46 Discoveries and Counting

First discovery: 2001el in NGC 1448
Number of discoveries: 46; all working alone
Location: 40 km East of Pretoria, South Africa
Equipment: 30 cm SCT

In terms of sheer numbers of supernova discoveries, the South African amateur Berto Monard comes fifth equal, after Tim Puckett, Michael Schwartz, Tom Boles, Mark Armstrong, and currently tying with the legendary visual discoverer, Bob Evans. He had made 46 discoveries of 14th to 18th magnitude supernovae up to September 2006. However, unlike the other four CCD observers, Berto, like Bob Evans, lives in the Southern Hemisphere. Originally from Belgium, Berto Monard has lived in Pretoria, South Africa, since 1981 and had an impressive record as an amateur astronomer even before he started discovering supernovae in September 2001. In the 4 years prior to that date, Berto was dedicated to studying Cataclysmic Variables, that is, unusual binary stars with regular brightness surges; they are not unlike the Type Ia supernova systems, except with recurrent, not catastrophic outbursts. Berto is a team member of the Centre for Backyard Astrophysics (CBA), which is a global network of small telescopes dedicated to the photometry of CVs. Between 1990 and 1997, he was a visual observer, notching up an impressive 29,000 magnitude estimates of variable stars. Berto's supernova discoveries, like those of Ron Arbour, have all been made with a humble 30-cm Meade LX200. His observatory and telescope are shown in Figures 10.16, 10.17, and 10.18. This Observatory

Figure 10.16. Berto Monard stands next to his "Bronberg" Observatory, 40 km east of Pretoria. A 30-cm LX200 is used for patrolling and a Dobsonian is used for visual observing. Image: Berto Monard.

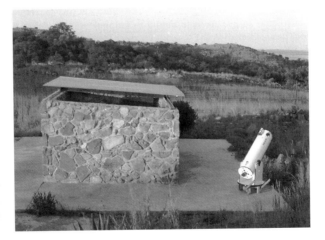

Figure 10.17. Another view of Berto Monard's observatory, with the roof open. Image: Berto Monard.

Figure 10.18. Berto Monard's LX200 telescope inside his run-off roof observatory. Image: Berto Monard.

(known as the Bronberg Observatory and also as CBA Pretoria) was established on the top of a 1,590-m-high ridge, 40 km east of Pretoria, in 2001. An SBIG ST7XME camera is used to take the images and a focal reducer (nominally 0.33×) gives a final f-ratio of 3.7. After his first discovery (2001el in NGC 1448), he went on to discover 6 supernovae in 2002, 4 in 2003, 10 in 2004, and a remarkable 14 in 2005. 11 more were bagged between January and July 2006. Arguably his most celebrated discovery was his first, of SN 2001ei. This was an unusual "asymmetric" Type Ia supernova and came after more than 12,000 negative galaxy patrols. It was caught on the rise and a lot of professional publications resulted from its study. Berto's discoveries 2005Q and 2005me occurred in the same galaxy (ESO 244-31, roughly 250 million light-years away) 11 months apart and 2005Q was still just faintly visible on the best images of 2005me.

Undoubtedly, living in the Southern Hemisphere has big advantages as there is simply far less competition that far south, apart, that is, from Bob Evans himself. More than half of Berto's finds have been further south than minus 30° declination. However, there have been major exceptions. His 17th mag discovery, SN 2005db in NGC 214, was at declination +25, although the short Northern Hemisphere summer nights of his rivals were working in his favor there. Another very memorable find was 2004gt, a 14th magnitude supernova in NGC 4038 (alias Caldwell 60) the northwestern component of the famous interacting galaxy pair called the Antennae. Many professional astronomers attempted to find the progenitor star for that supernova in previous exposures, and many postdetonation images have been secured, by large telescopes, to try to pin down its precise position. Some excellent images of some of Berto's most impressive supernovae, taken by Sergio Gonzales, are shown in Figures 10.19 to 10.24.

Figure 10.19. Supernova 2004ej in NGC 3095, discovered by Berto Monard. Image by Sergio Gonzales (Carnegie Supernova Project/Las Campanas Observatory) communicated by Berto Monard.

Figure 10.20. Supernova 2004ew in ESO 153-G17, discovered by Monard, Jacques, and Pimentel. Image by Sergio Gonzales (Carnegie Supernova Project/Las Campanas Observatory) communicated by Berto Monard.

Figure 10.21. Supernova 2004gt in NGC 4038 (alias Caldwell 60 of the Antennae), discovered by Berto Monard. Image by Sergio Gonzales (Carnegie Supernova Project/Las Campanas Observatory) communicated by Berto Monard.

Figure 10.22. Supernova 2005bf in MCG +00-27-5, discovered by Berto Monard and the Lick Observatory. Image by Sergio Gonzales (Carnegie Supernova Project/Las Campanas Observatory) communicated by Berto Monard.

Figure 10.23. Supernova 2005me in ESO 244-31, discovered by Berto Monard. Image by Sergio Gonzales (Carnegie Supernova Project/Las Campanas Observatory) communicated by Berto Monard.

Figure 10.24. Supernova 2005Q in ESO 244-G31, discovered by Berto Monard. Image by Sergio Gonzales (Carnegie Supernova Project/Las Campanas Observatory) communicated by Berto Monard.

The Top Patrollers' Favorite Galaxies

If patrolling thousands of galaxies is just too painful to contemplate, you might get lucky with a smaller sample-size consisting of the most productive galaxies. Of course, everyone else will be patrolling these freak galaxies, too, but someone has to strike it lucky. Of course, many of the most productive galaxies may well just be statistical flukes and many are simply big, nearby, face-on spirals whose supernovae will be easier to spot. However, in the case of a galaxy like NGC 6946 (see Figure 10.25), which has produced 8 supernovae in the past 90 years (and a few may well have been missed), there really does appear to be an inherent ability to produce far more supernovae than theory would predict. So these fertile supernova breeding grounds are definitely worth a regular check. Listed below are the 26 galaxies that have produced three or more supernovae in the past hundred years or so: a total of 94 supernovae all told! If these 26 galaxies regularly produce supernovae at these rates, then patrolling all 26 regularly (with no competitors) would yield a new discovery every year. Of course, in practice this collection of galaxies may well produce far less, or even far more than one supernova per year. But one thing you could bet on would be that you would be clouded out when the supernova appeared and someone else would bag it!

Figure 10.25. The positions of the eight supernovae discovered in NGC 6946 since the first in 1917. From far left to center: 1980k; 2004et; 1948B; 1968D. Three in a line: 1917A (top); 2002hh (middle); 1969P (lower); Far right: 1939C. Image/diagram by the author.

The Top Eight: Galaxies with Four or More Discovered Supernovae

NGC 6946 (Caldwell 12)
Constellation: Cepheus
20 h 34.9 m +60°09'
Supernovae: Eight from 1917 to 2006

Table 10.1. Supernovae in NGC 6946

Designation	Date	Discoverer	Disc. Mag.	Offset	Type
2004et	Sep 27	Moretti	12.8	250E 120S	II
2002hh	Oct 31	LOTOSS	16.5	61W 114S	II
1980k	Oct 28	Wild	11.4	280E 166S	II-L
1969P	Dec 11	Rosino	13.9	5W 180S	N/A
1968D	Feb 29	Wild & Dunlap	13.5	45E 20N	II
1948B	Jul 6	Mayall	14.9	222E 60N	II
1939C	Jul 17	Zwicky	13.0	215W 24N	N/A
1917A	Jul 19	Ritchey	14.6	37W 105S	N/A

NGC 5236 (Messier 83)
Constellation: Hydra
13 h 37.0 m −29°52'
Supernovae: Six from 1923 to 2006.

Table 10.2. Supernovae in Messier 83

Designation	Date	Discoverer	Disc. Mag.	Offset	Type
1983N	Jul 3	Bob Evans	12.5	120W 130S	Ia
1968L	Jul 17	Jack Bennett	11.9	5W 0S	N/A
1957D	Dec	Gates	15.0	41W 145N	N/A
1950B	Mar 15	Haro	14.5	105W 0N	N/A
1945B	Jul 13	Liller	14.2	97W 175S	N/A
1923A	May 5	Lampland	14.0	109E 58N	N/A

Note: The 1945B discovery was by Bill Liller, the modern day nova discoverer. He made the discovery in 1990 while examining archival plates taken at Harvard's Bloemfontein station.

NGC 4321 = Messier 100
Constellation: Coma Berenices
RA 12 h 22.9 m +15°49'
Supernovae: Five from 1901 to 2006

Table 10.3. Supernovae in Messier 100

Designation	Date	Discoverer	Disc. Mag.	Offset	Type
2006X	Feb 7	Suzuki	15.3	12W 48S	Ia
1979C	April 19	Johnson	12.1	56E 87S	II-L
1959E	Feb 21	Humason	17.5	58E 21S	N/A
1914A	Mar 2	Curtis	15.7	24E 111S	N/A
1901B	Mar 17	Curtis	15.6	110W 4N	N/A

NGC 2276

Constellation: Cepheus.

RA 7 h 27.2 m Dec +85° 45′

Supernovae: Five from 1962 to 2005.

Table 10.4. Supernovae in NGC 2276

Designation	Date	Discoverer	Disc. Mag.	Offset	Type
2005dl	Aug 25	Dimai & Migliardi	17.1	18E 1S	II
1993X	Aug 22	Treffers et al.	16.3	30E 69N	II
1968W	Mar 24	Iskudarian	16.6	7W 7N	N/A
1968V	Jan 26	Shachbazian	15.7	35W 36N	N/A
1962Q	Feb 25	Shachbazian	16.9	34W 11S	N/A

NGC 2841

Constellation: Ursa Major.

RA 9 h 22.0 m Dec +50° 59′

Supernovae: Four from 1912 to 1999

Table 10.5. Supernovae in NGC 2841

Designation	Date	Discoverer	Disc. Mag.	Offset	Type
1999by	April 30	Arbour et al.	15.0	96W 91N	Ia-p
1972R	Dec 5	Wild	16.0	46W 70S	N/A
1957A	Feb 26	Schurer	14.0	106W 73N	N/A
1912A	Feb 19	Pease & Curtis	13.0	50W 20N	N/A

NGC 3184

Constellation: Ursa Major.

RA 10 h 18.3 m Dec +41° 25′

Supernovae: Four from 1921 to 1999

Table 10.6. Supernovae in NGC 3184

Designation	Date	Discoverer	Disc. Mag.	Offset	Type
1999gi	Dec 9	Kushida	14.5	4W 61N	II
1937F	Dec 9	Zwicky	13.5	5E 149S	N/A
1921C	Dec 5	Jones	11.0	79E 236S	N/A
1921B	Apr 6	Zwicky	13.5	32E 160S	N/A

NGC 3690

Constellation: Ursa Major.

RA 11 h 28.5 m Dec +58° 34′

Supernovae: Four from 1992 to 1999

Table 10.7. Supernovae in NGC 3690

Designation	Date	Discoverer	Disc. Mag.	Offset	Type
1999D	Jan 16	Beijing Obs.	15.6	19W 5S	II
1998T	March 2	Beijing Obs.	15.4	N/A	Ib
1993G	March 5	Treffers et al.	16.6	2W 15S	II
1992bu	March 9	Van Buren et al.	16.6	5E 3S	N/A

NGC 4303 = Messier 61

Constellation: Virgo.

RA 12 h 21.9 m Dec +4 28'

Supernovae: Four from 1926 to 1999

Table 10.8. Supernovae in Messier 61

Designation	Date	Discoverer	Disc. Mag.	Offset	Type
1999gn	Dec 17	Dimai	16.0	32E 40S	II
1964F	June	Rosino	14.0	1S 14E	II
1961I	June 3	Humason	13.0	82E 12S	II
1926A	May 9	Wolf & Reinmuth	14.0	11W 69N	N/A

There are also 18 further galaxies that have produced at least three supernovae, each within the past hundred or so years, namely (in NGC order): NGC 664; NGC 1097; NGC 1084; NGC 1448; NGC 2207; NGC 1365; NGC 3367; NGC 3627 (M 66); NGC 3631; NGC 4157; NGC 4254 (M 99); NGC 4374 (M 84); NGC 4725; NGC 5033; NGC 5457 (M 101); NGC 5468; NGC 6754; UGC 1993.

Almost 100 galaxies have had two observed supernovae in the past 100 years, and this number is increasing every year.

Two Supernovae at Once

Although double supernova galaxies are not that rare, galaxies exhibiting two bright supernovae simultaneously are *much* rarer. Remember, most supernovae brighten and then fade away in 6 months (unless they are in a nearby galaxy or you have access to a huge telescope). Nevertheless, as recently as September 2005, two bright supernovae were discovered within a couple of days of each other in the galaxy UGC 4132. Supernovae 2005en and 2005eo were spotted independently by Tim Puckett and Mike Peoples and by the Lick Observatory Supernova Search team. They were mag 17.5 and 18.3, respectively, at that time. It is important to realize that galaxies can be more than 100,000 light-years across, so light can take 100,000 years to cross them. Thus, in reality, two supernovae visible in the same galaxy at the same time might have actually occurred 100,000 years apart!

In 2003, the galaxy NGC 772 threw up a great spectacle for amateur CCD imagers (see Figure 10.26). On August 20, the Lick and Tenagra joint patrol (LOTOSS) had discovered a magnitude 16.5 Type II supernova (2003hl) 24 arc-seconds east and 13 arc-seconds south of the galaxy's nucleus. Seven weeks later, amateur astronomer J. L. Lapasset imaged the new supernova and discovered a second bright supernova in the field! SN 2003iq was 5 arc-seconds east and 46 arc-seconds south of the nucleus and was also magnitude 16.5. I was one of many amateur astronomers who imaged this rare spectacle of two really bright supernovae in one galaxy. 2002ha and 2003dt in NGC 6962 just qualify, although the former had faded to mag 19 when the latter was discovered at 17th mag. Supernovae 2002cr and 2002ed in NGC 5468 provided the same discovery opportunity for Berto Monard that Llapasset would have a year later (i.e., he was observing 2002cr when 2002ed erupted). In addition, Berto's discoveries 2005Q and 2005me occurred in the same

Figure 10.26. The rare occurrence of two bright supernovae visible in the same galaxy. Supernovae 2003hl and 2003iq, October 26, 2003. Image by the author using a Paramount ME/Celestron 14.

galaxy (ESO 244-31, roughly 250 million light-years away) 11 months apart, and 2005Q was still just faintly visible on the best images of 2005me.

At least 10 such examples are known, but, again, it all depends where you draw the line on the earlier supernova having faded to obscurity. Certainly, the 2003 NGC 772 example is the finest I can recall. Another example I have already mentioned is Bob Evans' second discovery, namely 1981D in NGC 1316. He discovered it at magnitude 12.7, less than 3 months after an earlier 12th magnitude supernova, 1980N, in the same galaxy, which was still obvious on photographs.

Chapter 11

Searching the Messier Galaxies

There are 39 Messier galaxies in total. Sometimes a 40th, M 102, is added in the form of NGC 5866 but this was not on Messier's original list. M 102 is usually assumed to be identical to M 101. The Messier galaxies are bright, which is why they could be spotted with primitive 18th century telescopes. With the equipment now in many amateurs' hands, they can easily be checked visually for supernovae. The majority of the Messier galaxies are best placed in spring for Northern Hemisphere observers. While hunting down as many Messier objects as possible (all of them in one night is achievable) is a favorite March-time occupation of deep sky fanatics, just finding all of the Messier galaxies in Virgo and Coma Berenices is enough of a challenge for most beginners. Of course, if you use an accurate GO TO system, the task is not especially formidable, but, for the beginner, with a 150-mm telescope, just seeing the fuzzy patch with certainty is something to be proud of. In practice, it is unfriendly back-breaking eyepiece positions, the freezing cold, and the infuriations of dew, cloud, light pollution, and equipment failures that sap the strength of most beginners when faced with such a challenge. Sitting indoors in the warm is simply far more pleasant. With an equatorially mounted telescope featuring good old-fashioned mechanical setting circles, there is a very good starting point to your galaxy quest (see Figure 11.1). Find the star Denebola, the most easterly bright star in the constellation of Leo, and set the right ascension circle to 11 hours 49 minutes. The declination circle should read $+14^1/_2°$. Then, simply swing the telescope due east by 30 minutes ($7^1/_2$ degrees) or to an R.A. of 12 hours 19 minutes. For a Dobsonian, without setting circles, just hop about two finder telescope fields east of Denebola. If nothing appears in the eyepiece, just move the telescope about by a field diameter or two. Suddenly, you should spot a diffuse misty patch. If you got my instructions right, you are now looking at the galaxy M 99: a convenient place to start. Thirteen other Messier galaxies are within 7 degrees of M 99. A further three, M 49 and M 61 to the south and M 64 (the Black-Eye Galaxy) to the northeast, are within 10 degrees.

Once you have found M 99, a simple chart should enable you to locate the other 17 Coma/Virgo Messier galaxies shown in Table 11.1 and in Figures 11.2, 11.3, and 11.4. If you are a Northern Hemisphere observer with a fantastic southern horizon, you can bag the huge and superproductive galaxy M 83 in Hydra in springtime, too. I have lumped it in with the Coma/Virgo/Canes Venatica/Ursa Major galaxies because it lies at a similar right ascension to them. However, at a declination of almost −30 degrees, it will never even reach 9 degrees altitude from the latitude of London. It is a tribute to both Messier, and the galaxy's size and brightness, that he bagged it from the latitude of Paris. An image of M 83 is shown in Figure 11.5.

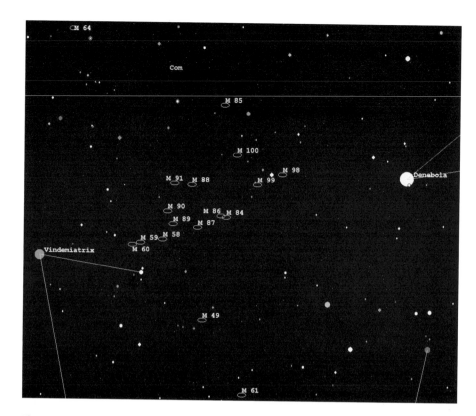

Figure 11.1. Messier galaxies close to the Virgo/Coma Berenices border. The field is approximately 20 degrees wide. Diagram by the author.

Table 11.1. The 18 Messier Galaxies in Virgo and Coma Berenices Plus M83 in the Far South

Galaxy	NGC	Mag	Approx. Size	RA	Dec
M 49	4472	8.4	10′ × 8′	12h30m	+8°00′
M 58	4579	9.8	6′ × 5′	12h38m	+11°49′
M 59	4621	10.6	5′ × 4′	12h42m	+11°39′
M 60	4649	8.8	8′ × 6′	12h44m	+11°33′
M 61	4303	9.7	7′ × 6′	12h22m	+4°28′
M 64	4826	8.5	10′ × 5′	12h57m	+21°41′
M 84	4374	9.3	7′ × 6′	12h25m	+12°53′
M 85	4382	9.2	7′ × 6′	12h25m	+18°11′
M 86	4406	9.8	10′ × 6′	12h26m	+12°57′
M 87	4486	9.6	9′ × 7′	12h31m	+12°23′
M 88	4501	9.5	7′ × 4′	12h32m	+14°25′
M 89	4552	9.8	5′ × 5′	12h36m	+12°33′
M 90	4569	10.3	10′ × 4′	12h37m	+13°10′
M 91	4548	11.1	5′ × 4′	12h35m	+14°30′
M 98	4192	10.8	9′ × 2′	12h14m	+14°54′
M 99	4254	9.8	5′ × 5′	12h19m	+14°25′
M 100	4321	10.2	8′ × 6′	12h23m	+15°49′
M 104	4594	8.3	9′ × 4′	12h40m	−11°37′
M 83	5236	8.1	13′ × 12′	13h37m	−29°52′

Figure 11.2. Messier galaxies M 49, M 58, M 59, M 60, M 61, and M 64 (from top left to bottom right).
All galaxy images have north at the top. All images by the author.

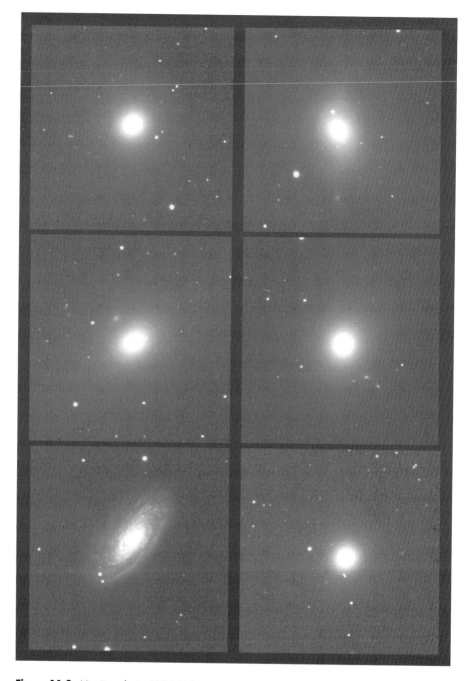

Figure 11.3. Messier galaxies M 84, M 85, M 86, M 87, M 88, and M 89 (from top left to bottom right). All images by the author.

Figure 11.4. Messier galaxies M 90, M 91, M 98, M 99, M 100, and M 104 (from top left to bottom right). All images by the author, except M 104, which was taken by Ron Arbour.

Figure 11.5. The most supernova productive Messier galaxy, M 83, captured here from Kitt Peak with a 50-cm f/8.4 RC Optical Systems telescope and an SBIG ST10XME CCD. The luminance (unfiltered) image was a 90-minute exposure. Color was provided from three, 20-minute, R, G, and B exposures (pixels binned 2 × 2). Image: Allan Cook/Adam Block/NOAO/AURA/NSF.

Figure 11.6. Messier galaxies in Canes Venatica, Ursa Major, and Draco. Messier 97, near M 108, is known as the Owl Nebula and is not a galaxy. Diagram by the author.

But in the Virgo, Coma Berenices region, any telescope larger than 150-mm aperture is capable of revealing countless dozens of NGC (New General Catalogue) galaxies unknown to Charles Messier.

Should you easily bag all of the Coma/Virgo border Messier galaxies, bear in mind that the whole strip of right ascension between 11 hours and (roughly) 14 hours has more to offer. Twenty degrees below the main Coma/Virgo Messier cluster and 11 degrees west of brilliant Spica you will find the so-called Sombrero Galaxy, M 104, right on the southern Virgo border. Moving 20 degrees in the opposite direction (i.e., north), past the star Cor Caroli and into central Canes Venatica and Ursa Major (Figure 11.6), we can bag the galaxies M 63 (the sunflower), M 94,

M 106, the famous and beautiful Whirlpool Galaxy, M 51 (all four feature in Figure 11.7), and on to M 108 and the lovely barred spiral M 109 (both shown in Figure 11.8). Back at the other end of the Bear, the large spiral M 101 sits just above the tail but will overspill many telescope/CCD fields of view when imaged (Figure 11.9). The bright galaxy often named M 102 (i.e., NGC 5866), although not actually in Messier's original catalogue, lies just over the Ursa Major border in Draco (Figure 11.10). Proceed north to almost +70 degrees Dec and M 81 and M 82 (Figures 11.11 and 11.12) are easy pickings.

Figure 11.7. Messier galaxies M 51, M 106, M 63, and M 94 (top left to bottom right). All images by the author.

Figure 11.8. Messier galaxies M 108 (left) and M 109 (right). Images by the author.

Figure 11.9. Most of the Messier galaxy M 101 (the outer edges are outside the field). Image by the author.

Figure 11.10. NGC 5866 is sometimes called M 102, even though it is almost certainly not the 102nd Messier object. Image: DSS/STScI.

Figure 11.11. Messier 81 in Ursa Major. Image: Gordon Rogers.

Figure 11.12. Messier 82 in Ursa Major. Image: Gordon Rogers.

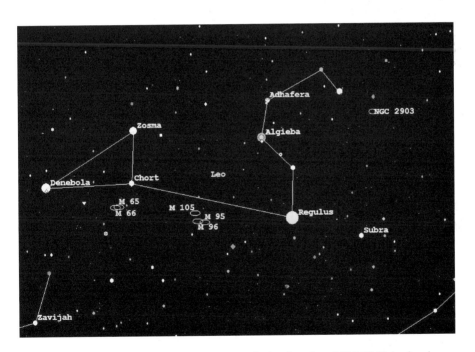

Figure 11.13. The five Messier galaxies in Leo. The bright galaxy NGC 2903 is also shown. Diagram by the author.

You are never short of galaxies when it is spring in the Northern Hemisphere. This whole longitude of the sky is literally teeming with galaxies and even beneath Leo (Figure 11.13) you will find another five from Messier's catalogue. M 65 and M 66 are only a few degrees below the bright star Chort and M 95, M 96, and M 105 are directly under a line connecting Chort to Regulus. All five are shown in Figure 11.14.

Figure 11.14. Messier galaxies M 65, M 66, M 95, M 96, and M 105 (top left to bottom right). Images by the author.

Table 11.2. The Nine Messier Galaxies in Canes Venatica and Ursa Major Plus the Galaxy M 102 (Not in the Original Catalogue)

Galaxy	NGC	Mag	Approx. Size	RA	Dec
M 51	5194	9.2	11′ × 7′	13h30m	+47°12′
M 63	5055	9.6	13′ × 7′	13h16m	+42°2′
M 81	3031	7.9	27′ × 14′	9h56m	+69°4′
M 82	3034	9.2	11′ × 4′	9h56m	+69°41′
M 94	4736	8.7	11′ × 9′	12h51m	+41°7′
M 101	5457	7.9	29′ × 27′	14h3m	+54°21′
(M 102)	5866	10.0	3′ × 1′	15h7m	+55°46′
M 106	4258	9.3	19′ × 7′	12h19m	+47°18′
M 108	3556	10.7	9′ × 2′	11h12m	+55°40′
M 109	3992	10.6	8′ × 5′	11h58m	+53°23′

Table 11.3. The Five Messier Galaxies in Leo

Galaxy	NGC	Mag	Approx. Size	RA	Dec
M 65	3623	10.1	10′ × 3′	11h19m	+13°6′
M 66	3627	9.6	9′ × 4′	11h20m	+13°0′
M 95	3351	10.7	7′ × 5′	10h44m	+11°42′
M 96	3368	9.9	8′ × 5′	10h47m	+11°49′
M 105	3379	10.1	5′ × 5′	10h48m	+12°35′

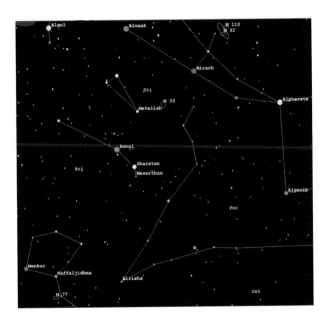

Figure 11.15. The Messier galaxies in Andromeda, Triangulum, Pisces, and Cetus. Diagram by the author.

Only six Messier galaxies lie outside the Northern Hemisphere's spring sky (see Figure 11.15); specifically they are at Northern Hemisphere autumnal RAs of between 0 and 3 hours. Few in number they may be, but two of them are absolute stonkers and there would be mayhem if they produced supernovae in the modern era, as they would be easy binocular targets. I am, of course talking about the great Andromeda Galaxy M 31 and M 33 in Triangulum (Figures 11.16 and 11.18). Two of the other galaxies, M 32 and M 110 (shown in Figure 11.17), are merely satellite galaxies of M 31. The other two Messier galaxies in that part of the sky are M 74 (a beautiful face-on spiral shown in Figure 11.19) and M 77 in Cetus, the Whale (shown in Figure 11.20). The whole Andromeda, Pegasus, Triangulum region is swarming with faint galaxies if you have a big telescope and a CCD camera, but the six Messier galaxies are, obviously, amongst the easiest to locate.

Figure 11.16. The Andromeda Galaxy, Messier 31, and
its companion galaxies M 32
(just below M 31) and M 110
(top right). Image: Ian Sharp.

Figure 11.17. The companion galaxies of Messier 31: M 32 (left) and M 110 (right).

Figure 11.18. The large face-on spiral M 33 in Triangulum. Image: Gordon Rogers.

Figure 11.19. Messier 74 and SN 2003gd. Imaged by the author.

Figure 11.20. Messier 77. Imaged by the author.

Table 11.4. The Six Messier Galaxies in Andromeda, Triangulum, Pisces, and Cetus

Galaxy	NGC	Mag	Approx. Size	RA	Dec
M 31	224	4.0	190′ × 60′	00h43m	+41°16′
M 32	221	9.1	9′ × 7′	00h43m	+40°52′
M 110	205	8.9	22′ × 11′	00h40m	+41°41′
M 33	598	6.2	70′ × 42′	01h34m	+30°40′
M 74	628	10.0	10′ × 10′	01h37m	+15°47′
M 77	1068	9.6	7′ × 6′	02h43m	−0°01′

Supernovae in the Messier Galaxies

So, of our 39 Messier galaxies, 40 if we include M 102, how many have actually produced supernovae that have been detected here on Earth? Well, so far, M 102 has produced none, so we can eliminate this infiltrator from our investigation straight away. Of the 39 genuine Messier galaxies, 24 have produced supernovae since 1885A in the Andromeda Galaxy (M 31). In fact, those 24 galaxies have actually produced a total of 46 supernovae, not least due to the phenomenal output of southerly M 83 (6 supernovae) and the following six galaxies: M 61 (4); M 100 (4); M 66 (3); M 84 (3); M 99 (3); M 101 (3). Even as I started writing this book, the fourth supernova in M 100 was being discovered (Figure 11.21). A further three galaxies, namely M 51, M 58, and M 74, have all produced two supernovae each. Interestingly, all six of these latter examples have occurred since 1988. I think this emphasizes both the increased patrol coverage by the professionals, especially after the Second World War, and the virtually total coverage of bright Messier galaxies since the 1980s and the advent of the first CCD patrols.

A complete table of the known Messier supernovae is shown in Table 11.5.

Figure 11.21. Supernova 2006X in Messier 100. Imaged by the author.

Table 11.5. Messier Galaxy Supernovae up to 2006

Galaxy	Supernovae	Amateur Messier discoveries
M 31	1885A	
M 49	1969Q (Not proved to be a supernova)	
M 51	1994I; 2005cs	Puckett et al. (94I); Kloehr (05cs)
M 58	1988A; 1989M	Ikeya & Evans (88A)
M 59	1939B	
M 60	2004W	
M 61	1926A; 1961I; 1964F; 1999gn	
M 63	1971I	
M 66	1973R; 1989B; 1997bs	Evans (89B)
M 74	2002ap; 2003gd	Hirose (02ap); Evans (03gd)
M 81	1993J	Garcia
M 82	1986D (probably not a SN); 2004am	
M 83	1923A; 1945B; 1950B; 1957D; 1968L; 1983N	Bennett (68L); Evans (83N)
M 84	1957B; 1980I; 1991bg	Romano (57B); Kushida (91bg)
M 85	1960R	
M 87	1919A	
M 88	1999cl	
M 96	1998bu	Villi
M 99	1967H; 1972Q; 1986I	
M 100	1901B; 1914A; 1959E; 1979C; 2006X	Johnson (79C); Suzuki (06X)
M 101	1909A; 1951H; 1970G	
M 106	1981K	
M 108	1969B	
M 109	1956A	

If we look at the 24 supernova producing Messier galaxies and the 15 that have failed to produce any in the past century, can we spot any obvious reason for this, or is it just a statistical quirk? Certainly when we look at a superb, and almost face-on, nearby galaxy like M 33 and compare it to, say, M 101, there would seem to be no obvious reason, other than random chance, that the former has produced no supernovae and the latter has produced three. Of course, supernovae can occur when a galaxy is too close to the sun to observe and, in that respect, M 101 and its Ursa Major companions have a distinct advantage. Productive NGC 6946 has the same advantage. From high Northern Hemisphere latitudes, high-declination galaxies will be circumpolar: visible all night, for every night of the year. However, the opposite is true for M 83, which is also a distinctly Southern Hemisphere–favored galaxy, with less observatories patrolling it; and yet six supernovae are associated with it. What can we make of all this?

Well, there is a whole mix of probabilities to consider, including:

- Observability throughout the year
- Whether the galaxy is presented face-on
- Whether supernovae are lost in the dazzling central bulge
- The number of stars in the galaxy
- The proportion of massive stars in the galaxy
- Distance to the galaxy (nearer supernovae will be brighter and spotted more easily)

- How well the galaxy was monitored in the early 20th century
- Random chance!

   If we consider the most edge-on Messier galaxies like M 82, M 98, M 104, and M 108, we can see that neither M 98 nor M 104 have ever been credited with a supernova, and of the two listed for M 82, 1986D was probably an HII region and 2004am was a very faint 17th mag object at discovery. 1969B in M 108 was a genuine bright(ish) object, though. While supernovae can occur in any galaxy at any time, the bright, big, nearby, face-on spirals that have proved productive in the past are the best to patrol if you cannot face imaging a thousand galaxies per night. In addition, it is interesting how many galaxies have produced more than one supernova in recent years, despite having produced none prior to the CCD era. This seems to indicate that we may be underestimating the supernova-producing capacity of some galaxies if we consider their output over the whole of the past century and do not allow for discoveries that may have been missed. Amateur patrollers quite routinely patrol in twilight and full moonlight these days, something that would not have been contemplated in the photographic era. Just look, for proof, at the Messier galaxies M 51, M 58, and M 74, which produced two supernovae each in the years 1994 and 2005 (M 51), 1988 and 1989 (M 58), and 2002 and 2003 (M 74), but no others. Five of these six supernovae were discovered by amateurs.
   Despite being a superb face-on example of a spiral galaxy, Messier 74 had not produced any detected supernovae until 2002, when it promptly produced a very special one indeed. SN 2002ap was discovered on January 29 of that year at magnitude 14.5, and it was spotted at a huge angular elongation from the galaxy. The supernova was more than 4 arc-minutes west and nearly 2 arc-minutes south of the nucleus, well outside the obvious spiral form visible in amateur images. The discoverer was Yoji Hirose, a Japanese amateur astronomer based at Chigasaki city, some 50 km west of Tokyo. Remarkably, his discovery was made on the day after full moon, with the sky awash with moonlight. This reminds me of a sage comment once made to me by Mark Armstrong, when he stated, just after a discovery: "There was thin cloud everywhere and a full moon in the sky – a typical discovery night!" Hirose used a 25-cm f/10 Schmidt-Cassegrain at f/6.3, together with an SBIG ST9 CCD camera for his SN 2002ap discovery. This combination gave a nice wide field of view of 22 arc-minutes, with an image scale of 2.6 arc-seconds per pixel. The telescope was mounted on a Takahashi EM 200 mounting. He had been searching for 22 years with five different telescopes before making his discovery! As described by Robin Scagell and Guy Hurst in the February 2002 issue of *The Astronomer* magazine, a day later Ostensen obtained a spectrum of the new supernova with the 4.2-m William Herschel telescope on La Palma. This was subsequently examined by supernova experts Peter Meikle and Stephen Smartt who noticed that the spectrum showed unusual lines, similar to those seen in the supernova 1998bw. That earlier supernova was linked to the gamma ray burst GRB 980425. SN 2002ap was designated a hypernova, although not the brightest example. Nevertheless, it brightened to magnitude 12.5 within 2 weeks of discovery and became one of the best studied supernovae in recent years. Perhaps the moral of this story is that even if you have been searching for 22 years, do not give up. The big one could be just around the corner. Remarkably, despite 2002ap being the first supernova found in M 74, the amazing Bob Evans found another supernova in that same galaxy less than 17 months later, on June 12, 2003. Yet again, the supernova was at a healthy elongation, although, this time it was obviously on a spiral arm and was not a hypernova.

Chapter 12

Searching the Caldwell Galaxies

Not everyone likes the term *Caldwell catalogue*. It was a catalogue devised in 1995 by the British amateur astronomer, prolific author, and TV personality Sir Patrick Moore in an attempt to list all of the "easy" deep sky objects not catalogued by Charles Messier in his 18th century catalogue. Although the Messier catalogue objects are some of the easiest for Northern Hemisphere astronomers to hunt down, the catalogue is not exhaustive and there are quite a few objects that Messier appears to have overlooked. In addition, there are quite a few that would have been difficult, or impossible, to observe from his Paris site. Thus, Patrick decided to emulate Messier's original 109-object catalogue and list his own favorite 109 objects that were *not* included in Messier's list. The name Caldwell was used because "Moore" would mean an "M" designation and Messier had already used "M." Patrick's surname is actually double-barrelled (i.e., Caldwell-Moore), so he used the Caldwell part and his objects have a "C" designation. There was never any claim that Patrick had discovered these objects, which seems to be the unfair comment that a few people have made regarding the name. They already have mainly NGC (New General Catalogue) designations. Patrick was just extending the list of easy-to-see deep sky objects beyond Messier, and the Caldwell catalogue has actually proved to be highly popular; even telescope hand controllers now feature keypads with Caldwell designations. Caldwell objects are listed in order of decreasing declination, which is far more useful than the Messier designations. At a stroke, you can draw a line under a certain number and say that all the objects with a higher number are too far south (or with a lower number are too far north for a Southern Hemisphere observer). In my case, the critical number, for a Caldwell galaxy, is probably 65 or NGC 253. At a declination of $-25°17'$ it can never rise more than 12 degrees above my south horizon. But it is a big, bright galaxy so I could probably get an image of it. I will have to try!

There are 35 galaxies in the Caldwell catalogue, (see Table 12.1, p. 189) and 16 of them have produced supernovae during the past century (see Table 12.2, p. 190); a somewhat poorer ratio than with the Messier galaxies, as would be expected from a sample containing more Southern Hemisphere objects. Nevertheless, single-handedly the awesome Caldwell 12 (=NGC 6946) makes up for this with its ridiculous total of eight supernovae!

The first four Caldwell galaxies [i.e., C3 (NGC 4236); C5 (IC 342); C7 (NGC 2403) and the aforementioned C12 (NGC 6946)] live at declinations of +60 degrees and above. They are shown in Figures 12.1, 12.2, and 12.3. Caldwell's 17 and 18 (Figure 12.4) alias NGC's 147 and 185 are large 9th magnitude galaxies that are only 1 degree apart in Cassiopeia. Caldwell's 21 and 23 (NGC 4449 and 891) lie at similar

Figure 12.1. Caldwell 3 = NGC 4236. Image: Tom Boles.

Figure 12.2. Caldwell 5 = IC 342. Image: Gordon Rogers

Figure 12.3. Caldwell 7 = NGC 2403 (upper), imaged by Jeremy Shears, and Caldwell 12 = NGC 6946 (lower), imaged by the author. Supernovae 2004dj and 2004et, respectively, are indicated.

declinations but could not be more different in appearance. NGC 4449 is an irregular sprawling mess of a galaxy, whereas NGC 891 is a sharply defined edge-on splinter (see Figures 12.5 and 12.6). Caldwell 24 (Figure 12.7), or NGC 1275, is just the brightest fuzzy galaxy of many in the Perseus cluster.

Figure 12.4. NGC 185 (left) and 147 (right). Image by kind permission of Bill Patterson of Los Angeles, California (www.laastro.com).

Figure 12.5. Caldwell 21 = NGC 4449. Image: Gordon Rogers.

Figure 12.6. Caldwell 23 = NGC 891. Image: Gordon Rogers.

Figure 12.7. Caldwell 24 = NGC 1275. Image: Tom Boles.

The Caldwell galaxies are not quite as numerous in the Canes Venatica/Coma Berenices region as the Messier galaxies, but they do still have a notable presence in that part of the sky. Caldwell's 26, 29, 32, 35, 36, and 38 are all in those northern constellations. They are shown in Figures 12.8, 12.9, 12.10, 12.11, 12.12, and 12.13. Looking a bit out of place in these Caldwell listings of Northern Hemisphere

Figure   12.8. Caldwell 26 = NGC 4244. Image: Ron Arbour.

Figure   12.9. Caldwell 29 = NGC 5005. Image: Tom Boles.

Figure   12.10. Caldwell 32 = NGC 4631. Image: Gordon Rogers.

Figure 12.11. Caldwell 35 = NGC 4889 is the brightest galaxy in this galaxy-filled field. Image: Ron Arbour.

Figure 12.12. Caldwell 36 = NGC 4559. Image: Gordon Rogers.

Figure 12.13. Caldwell 38 = NGC 4565. Image: Gordon Rogers.

springtime galaxies we find Caldwell 30, which is a fine spiral in the Northern Hemisphere autumn constellation of Pegasus (see Figure 12.14).

Moving further south in the Caldwell listings brings us to Caldwell 40, alias NGC 3626, a rather faint galaxy in Leo (see Figure 12.15). Two rather more distinctive galaxies, in the Northern Hemisphere autumn sky in Pegasus, are Caldwell 43 and Caldwell 44, alias the edge-on galaxy NGC 7814 and the superb face-on barred spiral NGC 7479. These are shown in Figures 12.16 and 12.17. Moving further

Figure 12.14. Caldwell 30 = NGC 7331. Image by the author.

Figure 12.15. Caldwell 40 = NGC 3626. Image by Tom Boles.

Figure 12.16. Caldwell 43 = NGC 7814. Image by Tom Boles.

Figure 12.17. Caldwell 44 = NGC 7479. Image by Gordon Rogers.

south, Caldwell 45, in Bootes (Figure 12.18) and Caldwell 48 in Cancer (Figure 12.19) are a couple of great springtime objects to track down. They are also known as NGC 5248 and NGC 2775. Caldwell 51, alias IC 1613, is an irregular dwarf galaxy in Cetus and is often regarded as the hardest Caldwell galaxy (or object) to spot.

Figure 12.18. Caldwell 45 = NGC 5248. Image by Tom Boles.

Figure 12.19. Caldwell 48 = NGC 2775. Image by Tom Boles.

It is nicknamed the Scarecrow because of its highly disorganized "scarecrow-like" appearance. It is also the last of the positive declination Caldwell galaxies (see Figure 12.20). When we get to Caldwell 52, or NGC 4697, we are in the Southern Hemisphere of the sky, although still within Virgo. NGC 4697 is a giant elliptical galaxy and an easy object in even modest apertures (see Figure 12.21). Moving ever southward we come to Caldwell 53, alias NGC 3115 or the Spindle Galaxy, well captured in the excellent image by Daniel Verschatse and an often overlooked galaxy

Figure 12.20. The irregular galaxy IC 1613 = Caldwell 51. Image: DSS/STScl.

Figure 12.21. NGC 4697 = Caldwell 52. Image: DSS/STScl.

(see Figure 12.22). Caldwell 57 (Figure 12.23) is a sprawling messy looking galaxy but is still within range of Northern Hemisphere patrollers at a declination of almost −15°. Below this Caldwell number the remaining galaxies are distinctly Southern Hemisphere patrol objects. The beautiful Antennae Galaxies of Caldwell's

Figure 12.22. A superb image of Caldwell 53 = NGC 3115. Image: Daniel Verschatse – Observatorio Antilhue, Chile (www.astro-surf.com/antilhue/).

Figure 12.23. The irregular galaxy NGC 6822 = Caldwell 57. Image: DSS/STScI.

60 and 61 have already produced one amateur find by Berto Monard of South Africa (see Chapter 10 and Figure 10.21 in that chapter). The remaining eight Caldwell galaxies from 62 to 101 are below −20 declination and I have not included figures of them here. However, they are well worth patrolling if you happen to live in the Southern Hemisphere. The Digitized Sky Survey Web page at http://archive.stsci.edu/cgi-bin/dss_form will quickly locate any of these galaxies and show a master image, once the NGC number is entered.

Apart from Caldwell 12/NGC 6946, the other most productive Caldwell galaxies are C 7, C 60, and C 67, namely NGC 2403, NGC 4038, and NGC 1097, respectively. NGC 2403 is a large but clumpy looking galaxy tilted only 28° from being edge-on and located in the rather dull constellation of Camelopardalis (the Giraffe). One of the brightest supernovae of all time was discovered in this galaxy as recently as July 31, 2004, by the Japanese amateur astronomer Itagaki. SN 2004dj was a

Table 12.1. The Caldwell Galaxies

Galaxy	NGC/IC	Mag	Size (arc-min)	RA	Dec
C 3	4236	9.6	20 × 8	12h 16.7m	+69°28′
C 5	IC 342	8.4	16 × 16	3h 46.8m	+68°06′
C 7	2403	8.4	24 × 13	7h 36.9m	+65°36′
C 12	6946	8.8	11 × 9	20h 34.9m	+60°09′
C 17	147	9.5	18 × 11	00h 33.2m	+48°30′
C 18	185	9.2	17 × 14	00h 39.0m	+48°20′
C 21	4449	9.6	5 × 4	12h 28.2m	+44°06′
C 23	891	9.9	12 × 3	02h 22.6m	+42°21′
C 24	1275	11.9	3 × 2	03h 19.8m	+41°31′
C 26	4244	10.4	16 × 2	12h 17.5m	+37°48′
C 29	5005	9.8	6 × 3	13h 10.9m	+37°03′
C 30	7331	9.5	10 × 5	22h 37.1m	+34°25′
C 32	4631	9.2	15 × 4	12h 42.1m	+32°32′
C 35	4889	11.5	3 × 2	13h 00.1m	+27°58′
C 36	4559	10.0	11 × 5	12h 36.0m	+27°58′
C 38	4565	9.6	16 × 2	12h 36.3m	+25°59′
C 40	3626	11.0	3 × 2	11h 20.1m	+18°21′
C 43	7814	10.6	6 × 2	00h 03.2m	+16°09′
C 44	7479	10.8	4 × 3	23h 05.0m	+12°19′
C 45	5248	10.3	6 × 5	13h 37.5m	+08°53′
C 48	2775	10.1	5 × 4	09h 10.3m	+07°02′
C 51	IC 1613	9.2	19 × 17	01h 04.8m	+02°07′
C 52	4697	9.2	7 × 5	12h 48.6m	−05°48′
C 53	3115	8.9	7 × 3	10h 05.2m	−07°43′
C 57	6822	8.8	15 × 15	19h 44.9m	−14°48′
C 60	4038	10.5	11 × 6	12h 01.9m	−18°52′
C 61	4039	10.3	10 × 4	12h 01.9m	−18°53′
C 62	247	9.2	22 × 7	00h 47.1m	−20°46′
C 65	253	7.6	26 × 6	00h 47.6m	−25°17′
C 67	1097	9.2	10 × 6	02h 46.3m	−30°16′
C 70	300	8.1	20 × 15	00h 54.9m	−37°41′
C 72	55	8.1	34 × 6	00h 15.1m	−39°13′
C 77	5128	6.7	17 × 13	13h 25.5m	−43°01′
C 83	4945	8.8	19 × 4	13h 05.4m	−49°28′
C 101	6744	8.6	21 × 13	19h 09.7m	−63°51′

Table 12.2. Caldwell Galaxy Supernovae (1917–2005)

Galaxy	Supernovae	Amateur Caldwell Discoveries
C7	2004dj; 2002kg; 1954j	Itagaki (04dj)
C12*	2004et; 2002hh; 1980k; 1969p; 1968d; 1948b; 1939c; 1917a	Moretti (04)
C23	1986j	
C24	1968A	
C29	1996ai	
C30	1959D	
C36	1941A	
C44	1990U	
C48	1993Z	
C53	1935B	
C60	2004gt; 1974E; 1921A	Monard (04gt)
C65	1940E	
C67	2003B; 1999eu; 1992bd	Evans (03B); Aoki (99eu)
C77	1986G	Centaurus A SN by Evans
C83	2005af	
C101	2005at	Monard codiscovery

* C12 alias NGC 6946 is, of course, the most SN productive galaxy with eight discoveries from 1917 to 2004 in various parts of that galaxy.

considerable distance (160″ east and 10″ north) from the galaxy's nucleus meaning that its decline from maximum brightness was well observed. Remarkably, only 2 years earlier another supernova was discovered in NGC 2403, the much fainter SN 2002 kg. Being situated at 65° north means that this galaxy is circumpolar, and therefore it can be patrolled all year round from northern Europe and the northern USA/Canada.

Caldwell 60, which produced supernovae 2004gt, 1974E, and 1921A, is actually just the northwestern component of a superb Southern Hemisphere spectacle called the Antennae: two galaxies colliding with each other. Not surprisingly the other, southeastern, component is designated as Caldwell 61. Their NGC numbers are 4038 and 4039. At declinations of almost −19°, the Antennae can be seen from latitudes as high as the United Kingdom but only as they transit in a spring sky and, even then, never more than 20 degrees above the horizon. Apart from the "Antennae" nickname, NGC 4038 and 4039 are also occasionally referred to as the Ringtail and Rattail galaxies. As previously mentioned, a picture of the antennae appeared in Figure 10.21.

Caldwell 67, the only other triple supernova Caldwell galaxy, is another Southern Hemisphere galaxy, and all three of its supernovae were discovered in an 11-year span from 1992 to 2003, two by amateur hunters Evans and Aoki.

Beyond Caldwell

Of course, it goes without saying that there are more galaxies worth patrolling than just the 39 or 40 Messier galaxies and the 35 Caldwell galaxies. We have already seen that the dedicated supernova patrollers have databases containing 10,000 galaxies. But the Messier and Caldwell objects are a great place to start and a manageable number. However, even after the Caldwell Catalogue was devised by Sir

Figure 12.24. Not a Messier galaxy and not a Caldwell galaxy either, but NGC 2903 in Leo is still a superb galaxy for CCD imagers. Image by the author with a Celestron 14 and SBIG ST9XE CCD.

Patrick Moore, there were still a number of leftover, obvious, bright galaxy omissions; objects that are visible even in small telescopes and spectacular in CCD images.

For example, one of my favorite bright galaxies is NGC 2903 in Leo, just west of the famous Sickle shape (Figure 12.24), and yet it is not in Messier's list or the Caldwell list. Then there are NGCs 4214, 4395, 4490, 4485, 4656 (the Hockey Stick) and 5033 in Canes Venatica. Don't overlook NGC 7640 in Andromeda either. NGC 1023 in Perseus is another fine galaxy and (deep breath) NGCs 2841, 2976, 3184, 3198, 3359, 3631, 3718, 3726, 3893, 3938, 4051, 4088, 4096, and 4605 are all bright galaxies in Ursa Major. NGC 3344 and 3486 in Leo Minor as well as NGC 3628 in Leo are well worth patrolling as are NGCs 4216, 4535, and 4536 in Virgo. NGC 4274 and 4725 in Coma Berenices are often overlooked, and the thin sliver NGC 5907 in Draco is a beautiful sight in any large amateur telescope. We have already encountered NGC 772 in Aries; it displayed a double supernova in 2003. Don't neglect that one. NGC 2683 in Lynx is a fine target, too. If you live in the Southern Hemisphere, don't overlook NGC 134 and NGC 7793 in Sculptor, NGC 3621 in Hydra, or NGCs 134 and 5102 in Centaurus.

Chapter 13

Observing Supernova Remnants

Although SN 1987A is the only nearby supernova observed in the modern era, we have already seen that at least five closer supernovae have occurred in the past 1,000 years. Their remnants, as well as those of other suspected supernovae, have been studied in detail by professional astronomers. Indeed, throughout the night sky a wealth of supernova remnants (SNRs) are known. Dave Green, an astronomer at the Mullard Radio Astronomy Observatory in Cambridge, England, lists 235 galactic supernova remnants on his Web site at http://www.mrao.cam.ac.uk/surveys/snrs/.

Green's Catalogue of Galactic Supernova Remnants was originally published in Volume 32 of the *Bulletin of the Astronomical Society of India* in 2004 (when it then contained 231 entries). However, these remnants are, not surprisingly, mainly objects visible with professional radio telescopes and not amateur telescopes. The number of SNRs within amateur range is considerably smaller.

The Crab Nebula

Without doubt, the most famous supernova remnant in the sky is that of the Crab Nebula, also known as M 1, NGC 1952, or 3C 144 (see Figure 13.1). The "M" designation is, of course, Charles Messier's famous 18th century catalogue; NGC stands for the New General Catalogue of Clusters and Nebulae (new in 1888!); 3C is the designation for the third Cambridge catalogue of astronomical radio sources. M 1 is the remnant of the aforementioned supernova of AD 1054 that erupted near zeta Tauri on the eastern edge of Taurus and shone with twice the brightness of Venus. At 7,000 light-years away, the Crab's current visible size of some $6' \times 4'$ arcminutes corresponds with an actual size of 12×8 light-years. In other words, in the 950 years since the supernova exploded, material from the explosion has expanded across a distance greater than we are from our nearest star, Proxima Centauri. M 1 was actually discovered by John Bevis in 1731 but rediscovered independently by Messier in 1758. Of course, being 7,000 light-years away it actually went bang around 6000 BC. At the heart of the Crab there is still a remnant of the massive progenitor star, that is, a 16th mag star also known as pulsar NP 0532, which rotates at 30 times per second and pulses at X-ray, radio, and even optical wavelengths. Large-aperture telescopes that can image a 16th mag star in an exposure of 1/100 of a second have even produced movies showing this pulsar light-

Figure 13.1. The Crab Nebula, M 1, imaged by Ron Arbour.

house flashing at 30 times per second. To find M 1 you simply go to the 3rd magnitude star zeta Tauri and then move a degree to the northwest and it will be obvious in the eyepiece of any amateur telescope. Alternatively, M 1 appears on the keypad of almost every GO TO telescope. It is an obvious magnitude 8.4 misty patch in the telescope eyepiece and is even visible in 10 × 50 handheld binoculars.

The Veil Nebula

A much larger deep sky SNR target, visible to both the visual binocular and visual telescopic observer, is the Veil (sometimes called Bridal Veil) Nebula in Cygnus. If you go back prior to the 1970s, objects as ghostly as the Veil were almost considered "impossible" targets for the visual observer. The low surface brightness certainly makes it a challenging object, but the availability of high-contrast and emission line nebula filters largely pioneered by companies like Lumicon have helped pull the ghostly details out of the skyglow. There are a number of good deep sky filters on the market, but the best ones for SNR viewing are those that pass the two oxygen III emission lines at 496 and 501 nm, such as the Lumicon Oxygen III filter, the Celestron Oxygen III Narrowband filter, and the Meade Series 4000 Oxygen III filter. Ultrahigh contrast (UHC) filters that pass the O III lines and the hydrogen-beta line at 486 nm are excellent, too. However, those living in fairly light-polluted sites will prefer the narrower O III filters as they darken the background more and there are only a few nebulae that benefit from a hydrogen-beta window. All other objects apart from O III emitters will be severely dimmed by an O III filter, but the contrast enhancement on supernova remnants like the Veil is stunning. It is amazing that such targets were considered virtually impossible prior to the 1970s.

Another factor here is the increasing willingness of amateurs to transport powerful instruments to very dark sites and to indulge in communal events like "star parties". At such events there are always experienced observers, large apertures, and a range of nebula filters for all to see and use. However, before we get carried away with aperture fever, I would like to stress that you do *not* need a massive telescope to see the Veil Nebula. It is, perhaps surprisingly, a very large deep sky object whose two halves are part of a circle almost 3 degrees in diameter! This is not dissimilar in size to the famous Andromeda Galaxy, M 31. However, it is the surface brightness of the two arcs that is the challenge and not its size, hence the importance of dark skies and appropriate filters. The Veil Nebula's two halves (see Figures 13.2 and 13.3) are in a very familiar part of the summer sky for Northern Hemisphere observers. They are situated just a few degrees south of the easternmost star of Cygnus' famous five-star cross, namely epsilon or Gienah Cygni. Gienah is a magnitude 2.5 easy naked-eye star and is situated at a declination of +34°. Move just over 3 degrees south of Gienah and you will find the 4th magnitude star 52 Cygni. The Veil's western component, NGC 6960, actually passes through the field of this star, which is both a curse and a blessing. It makes the western component easy to locate, but you will be tempted to nudge the telescope so that 52 Cygni is outside the field, so the star does not dazzle your view. In a quality, wide-field telescope (for example, a 100-mm apochromat used at about 20×) you will just be able to fit the whole Veil in your field of view. With a wide field of just over 2.5° and 52 Cygni at the western edge of the field, the eastern section just fits in. The western arc has the designation NGC 6960 and the eastern arc is composed of NGC 6992–NGC 6995. The western and eastern arcs also have Caldwell designations of Caldwell 34 and 33, respectively. The Veil arc is not complete. There is a major chunk missing to the southeast, even in deep photographs and images, but it is a beautiful and delicate sight in a dark sky, especially if you have an oxygen III or

Figure 13.2. The NGC 6960 portion of the Veil Nebula, imaged by Jeremy Shears.

Figure 13.3. The NGC 6992 portion of the Veil Nebula, imaged by Gordon Rogers.

Ultrahigh contrast filter. The current thinking is that the Veil Nebula actually lies at a distance of roughly 1,500 light-years from Earth, based on observations made by the Hubble Space Telescope in 1999. (As an aside, this is more than seven times further than 52 Cygni, which is about 200 light-years from Earth). This 1,500 light-year distance makes the Veil five times closer than the Crab Nebula. In terms of its age in our night sky, though, the Veil is much older. The supernova responsible probably exploded in our skies around 5,000 to 8,000 years ago. However, it is not inconceivable that both the Crab and the Veil supernovae actually went bang at about the same time. Ignoring, for simplicity, the fact that, since Einstein's relativity theories, there has been no such thing as an absolute time clock, it is possible that both the Veil and the Crab supernovae detonated 8,000 years ago. But, the light from the Veil supernova would have arrived after 1,500 years and that from the Crab after 7,000 years. In a large galaxy, the finite speed of light can sometimes look very slow indeed and completely skews our perception of events. If the Veil supernova was only 1,500 light-years away, it may well have been a dazzling magnitude −9 object in the skies of our ancestors. A distance of 1,500 light-years implies that the almost 3-degree diameter it now covers corresponds with a diameter of almost 80 light-years!

The Jellyfish Nebula

Compared with the Crab Nebula, IC 443 in Gemini is an extremely difficult object to track down. Like the Veil and the Crab, it is located in a very obvious location because it is between two naked-eye stars, namely mag 2.9 Mu Geminorum (also called Tejat Posterior) and mag 3.3 Eta Geminorum (also called Praepes). Yet again, these stars can be helpful and a dazzling hindrance, too, especially in wide-field instruments. Nebulosity exists all the way between these two stars on photographs but the bright part (i.e., IC 443 itself) is less than a degree ENE of Praepes. In long exposures, the NE component of the supernova remnant (Figure 13.4) looks

Figure 13.4. IC 443, the Jellyfish Nebula, imaged by Gordon Rogers.

remarkably like a jellyfish! This nebula is a really tough challenge and unless you have a very dark site or some UHC or OIII filters, you will really struggle to find it visually in any amateur instrument, even though IC 443 is over half a degree across. With wide field, filtered CCD equipment, though, it is easily revealed.

Simeis 147

Another highly challenging Northern Hemisphere SNR is Simeis 147 (also known as Sh2-240). This is another SNR that responds well to CCD imaging with H-alpha filtered telephoto lenses or wide-field apochromats, showing intricate filamentary structure (see Figure 13.5 by one of the world's best astro-imaging experts, Robert Gendler). It was discovered as recently as 1952 at the Crimean Astrophysical Observatory in Simeis in the Ukraine, hence the unusual name. Located around RA 5h 40m and Dec +28°, like the Veil it occupies over 3 degrees of sky. If you found the region of the Crab without a problem, it is right next door and near to an even brighter star, namely El Nath or Beta Tauri, the bright (mag 1.7) naked-eye star that is literally shared between Taurus and Auriga. On older star charts it is labeled as Gamma Aurigae not Beta Tauri. Simeis 147 is roughly 3 degrees east and slightly south of Beta Tauri. I say roughly because of its sheer size. However, despite its physical size, a large-aperture telescope fitted with an oxygen III filter will probably be the best visual strategy with this SNR. The region at the southern edge is supposed to be the easiest to glimpse, at around 5h 38m and +26°50' but its components are very elusive objects. Simeis 147 is thought to lie at a distance of 3,000 light-years (i.e., midway in distance between the Crab and the Veil). However, it is

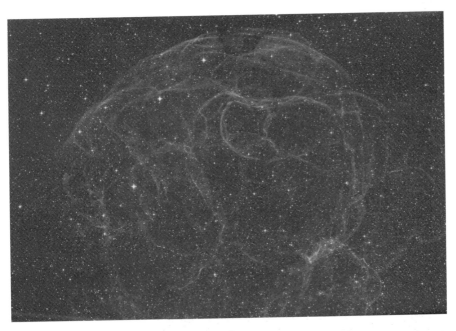

Figure 13.5. Simeis 147 imaged by deep sky expert Robert Gendler using a Takahashi FSQ 106 apochromat and SBIG ST11000 CCD. The luminance exposure was 8 hours with an H-alpha filter and the color was from three 30-minute exposures through red, green, and blue filters.

far, far older than either with an estimated age of 100,000 years. Thus its 3-degree radius corresponds with a diameter of 150 light-years, or 35 times the distance between us and the nearest star. Interestingly, there is a pulsar remnant of the supernova, spinning with a period of 143 milliseconds (i.e., seven times a second), and detected at radio and X-ray wavelengths. The pulsar is designated as J0538+2817, corresponding to its position of: RA 5 hr 39.1′ Dec +28°00′ (2000).

Sharpless 2-91/2-94

Another elusive and relatively recently designated supernova remnant is Sharpless 2-91/2-94, which, like the Veil, is also in Cygnus, not far from Albireo. Obviously, one expects the Milky Way constellations to bag the most supernova remnants but the Taurus/Auriga and Cygnus regions seem to have been favored. Alternative designations for this are LBN (Lynds Bright Nebulae) 147 and G65.3 +5.7. It was only discovered to be a galactic SNR when an emission line survey of the galactic plane revealed it in 1997. The brightest portion (i.e., Sh 2-91) itself lies roughly 15 arc-minutes south of Phi Cygni, which is a mag 4.7 naked-eye/binocular star (depending whether you live in the country or the town) roughly 2 degrees north of the famous double star Albireo. In filtered CCD images, a structure almost 4 degrees across is revealed. Visually the section close to 19 h 40 m +30° is the easiest to see in a large telescope equipped with an OIII filter.

The Crescent Nebula

Staying in the constellation of Cygnus, I have to mention the Crescent Nebula, NGC 6888, even though not all astronomers consider it to be a SNR (see Figure 13.6). If you continue moving north along the spine of the Swan, on a line from Albireo to Sadir, seven-eighths of the way to Sadir you will encounter the Crescent Nebula.

Figure 13.6. The Crescent Nebula NGC 6888 imaged by Jeremy Shears.

This is an emission nebula normally associated with the 7th mag Wolf-Rayet star HD 192163, also known as V1770 Cyg. The Crescent is, not surprisingly, crescent shaped, with dimensions of $20' \times 10'$ and a total visual magnitude of 10.0. Images from IRAS (Infra-Red Astronomy Satellite) have shown another nebula shell outside the crescent shape, and this could be the result of an old supernova explosion. However, many astronomers think this is just an earlier ejection from the Wolf-Rayet star. This may or may not be a SNR, but it is still a fascinating object to track down, visually or photographically. Again, the use of a nebular filter will help considerably.

The Witch Head

Another highly disputed SNR is IC 2118, the Witch Head Nebula in Eridanus; so-called because of its appearance in long exposure photographs. Different sources differ over this nebula's origin or just stick on the fence, calling it simply "a nebula," a "bright nebula," or an "emission nebula." The Witch Head is located only 3 degrees to the west of brilliant Rigel, just over the constellation border, in Orion, so, once again, finding the field is easy. Some sources state that it is a reflection nebula, reflecting the light of Rigel. It can be glimpsed in apertures as small as 100 mm.

The Vela Supernova Remnant

In the Southern Hemisphere, the Vela Supernova Remnant (Figure 13.7) makes up a section of the huge Gum Nebula (Gum 12), a complex shell of intricate ghostly filaments discovered by the Australian astronomer Colin Gum in 1952. The Gum is actually the largest object in the sky, apart from the Milky Way, and stretches more than 35 degrees across Vela and the surrounding areas. Within this nebulosity lies a long thin remnant, designated NGC 2736, which is aptly named the Pencil Nebula (Figure 13.8). This is a bright fragment of the 5-degree-diameter Vela Supernova Remnant on the eastern part of that huge bubble of expanding material. Many intricate veil-like filaments are seen on the western part of the nebula, but the bright thin Pencil is the most obvious visual feature on its eastern side. Much of the eastern curve is obscured by galactic dust, and the whole feature sits amongst a backdrop of myriads of Milky Way stars. John Herschel actually first observed the nebula called NGC 2736 in the 1840s. Once again, a very large telescope and an O III filter will be your best bet for a good view of the Pencil. The Vela supernova itself is thought to lie at a distance of only 800 light-years (815 is often quoted, although such accuracy is not warranted), that is, half the distance of even the Veil supernova and about nine times closer than the Crab supernova. Its age (i.e., how long ago it flared in our skies), is thought to be roughly 11,000 years. If the Vela supernova was this close, it would have shone as brightly as mag −10, turning night into twilight for months after its detonation for anyone in the Southern Hemisphere. Of course, a lot depends on which month it went off in and whether the supernova was circumpolar from our distant ancestors' location. It would have been a spectacular and awesome sight in the night sky. The entire Vela Supernova Remnant is now estimated to be about 114 light-years across (i.e., a radius of 57 light years from the supernova), and therefore the wispy material we

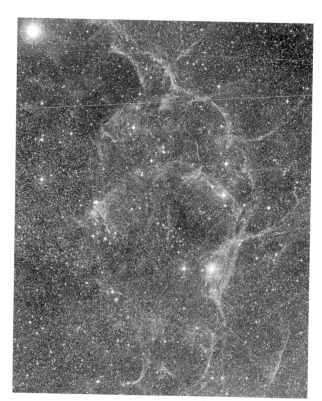

Figure 13.7. The giant Vela Supernova Remnant. This image is 2 degrees wide. Image: © 1987 Anglo-Australian Observatory. Photograph by David Malin. (see color plate)

Pencil Nebula • NGC 2736

Hubble
Heritage

NASA and The Hubble Heritage Team (STScI/AURA) • Hubble Space Telescope ACS • STScI-PRC03-16

Figure 13.8. The Pencil Nebula section of the Vela Supernova Remnant. Image: NASA/Hubble Heritage/STSCI/AURA.

Table 13.1. A Table of Some SNRs within Range of Amateur Telescopes (Visual or CCD)

Catalogue numbers	Pop. Name	RA	Dec	Dims	Size	Dist	Age	Con
M1, NGC 1952	Crab	5h 34.5m	+22°01'	6 × 4'	12 × 8 ly	7,000	953 yr	Tau
NGC 6960, 6992–95	Veil	~20h 50m	~+31°	~$2\frac{1}{2}$°	80	1,500	6,000?	Cyg
IC 443	Jellyfish	6h 18m	+22°45'	~40'	60	5,000	30,000	Gem
Simeis 147	Simeis 147	5h 40m	+28°	~3°	150	3,000	100,000	Tau
Sharpless 2-91	Sharpless 2-91	19h 40m	+30°	~4°	N/A	N/A	N/A	Cyg
NGC 6888	Crescent	20h 12m	+38°20'	~20'	30	5,000	N/A	Cyg
IC 2118	Witch Head	5h 4m	–7°	3° × 1°	40 × 15	~800	N/A	Eri
NGC 2736, etc.	Vela/Pencil	8h 34m	–46°	~8°	~114	800	11,000	Vel
Abell 85	Abell 85	23h 59m	+62°15'	35'	N/A	N/A	N/A	Cas

Dims stands for the object's dimensions in arc-minutes or degrees. Size is the estimated size in light-years or degrees. Distance is the distance, in light-years, to the object. Age is the estimated time since the supernova exploded in our skies. (so the size will be bigger now). Distance is the distance, in light-years, to the object. Age is the estimated time since the supernova exploded in our skies. Of course, the actual age will be longer by an amount equal to the distance. Thus the Crab Nebula is 953 years old in 2007 (i.e., 953 years since the supernova of 1054). However, in reality it is about 8,000 years old as the light took 7,000 years to reach us. Con is the constellation abbreviation. Tau = Taurus; Cyg = Cygnus; Gem = Gemini; Eri = Eridanus; Vel = Vela; Cas = Cassiopeia.

see has traveled one-fourteenth of the distance toward us in the past 11,000 years, at an average speed of roughly 57/11,000 times the speed of light (i.e., 1/200 of the speed of light or 1,500 kilometers/second). Remarkably, the Vela supernova pulsar, that is, the dense remains of the supernova's core, has been identified optically, despite only having a magnitude of 24. It was found in 1977 during a very sensitive search with the Anglo-Australian Telescope situated at Coonabarabran, New South Wales, Australia. It rotates 11 times per second and also emits X-rays although it does not actually pulse at those wavelengths. However, the Vela pulsar does give off regular gamma ray pulses, and it is actually the most intense source of gamma rays in the sky. A reminder, perhaps, of why we really do not want supernovae going off within 100 light-years of Earth!

Abell 85

Finally, I would like to return to the far northern skies and the constellation of Cassiopeia for another extremely elusive SNR, also known as Abell 85 or the radio source CTB1. It was originally catalogued as a planetary nebula by George Abell hence its presence in his catalogue. The brightest part of this SNR lies at roughly 23 h 59 m +62 d 15′, or 3 degrees north of 2nd magnitude Beta Cassiopeia (also called Caph) and 3 degrees east of the open cluster Messier 52. Being right in the Milky Way band, the whole region is awash with stars when viewed through binoculars or a telescope, making identification of this ghostly SNR a task for large apertures, an O III or H-beta filter, and very dark skies. Only the keenest deep sky experts will find this SNR visually, and even the filters only make the object slightly more obvious. In CCD images, a crescent-like arc over half a degree wide is seen stretching from east to west across the field. Table 13.1 lists all the vital data on these supernova remnants to allow easy comparison.

Appendix

Useful Supernova Data and Contacts

Relevant Astronomy, Supernova, and Galaxy Image Web Pages

Organizations

The Astronomer Supernova Pages: http://www.theastronomer.org/supernovae.html
The British Astronomical Association: http://britastro.org/baa/
American Association of Variable Star Observers: http://www.aavso.org/
Carnegie Supernova Project: http://csp1.lco.cl/~cspuser1/CSP.html
National Optical Astronomy Observatory Gallery http://www.noao.edu/image_gallery/

Amateur Astronomer's Supernova and Deep Sky Web Pages

Tom Boles Web site: http://www.coddenhamobservatories.org/
Dave Bishop's Supernova pages: http://www.rochesterastronomy.org/supernova.html
Robert Gendler: http://www.robgendlerastropics.com/
Gordon Rogers Deep Sky Web site: http://www.gordonrogers.co.uk/
Tim Puckett's site: http://www.cometwatch.com/search.html
Tenagra Observatories site: http://www.tenagraobservatories.com/
Bill Patterson: http://www.laastro.com
Daniel Verschatse: http://astrosurf.com/antilhue

Manufacturers Gallery Pages (for Galaxy Images)

SBIG gallery: www.sbig.com/sbwhtmls/gallery.htm
RCOS Messier gallery: http://gallery.rcopticalsystems.com/index.html#messier
RCOS NGC gallery: http://gallery.rcopticalsystems.com/index.html#ngc

Digitized Sky Survey Image Retrieval Page

http://stdatu.stsci.edu/cgi-bin/dss_form

Nearby Supernova Factory

http://snfactory.lbl.gov/

Katzmann Automatic Imaging Telescope (Lick Observatory)

http://astron.berkeley.edu/~bait/kait.html

SLOAN Digital Sky Survey

http://www.sdss.org/

CBAT/IAU Suspect Supernova Minor Planet Checker

http://scully.harvard.edu/~cgi/CheckSN

CBAT/IAU List of All Supernovae

http://cfa-www.harvard.edu/iau/lists/Supernovae.html

CBAT/IAU List of Recent Supernovae

http://cfa-www.harvard.edu/iau/lists/RecentSupernovae.html

Some Relevant Equipment Suppliers

SBIG (for CCDs and Spectrographs): http://www.sbig.com/
Starlight Xpress CCDs http://www.starlight-xpress.co.uk/
Software Bisque (*The Sky*, *CCDSoft*, and the Paramount ME): http://www.bisque.com/
Celestron International: http://www.celestron.com/
Meade: http://www.meade.com/
Rainbow Optics Spectroscopes: http://www.starspectroscope.com/
Guide 8.0 planetarium/telescope control software: http://www.projectpluto.com/

SBIG Spectrometer User Group

http://groups.yahoo.com/invite/SBIG-SGS

Blink Comparator Software

Dominic Ford's *Grepnova* software: http://www-jcsu.jesus.cam.ac.uk/~dcf21/astronomy. html

Galaxy Groups and Clusters Observing Guide

http://www.astroleague.org/al/obsclubs/galaxygroups/

Galaxy Triplets Web Page

http://www.angelfire.com/id/jsredshift/triplets.htm

Hickson Compact Galaxy Groups

http://www.angelfire.com/id/jsredshift/hickcatalog.htm

The Principal Galaxy Catalogues in Use by Supernova Patrollers

Messier: Charles Messier's 18th century catalogue, which included 39 of the brightest galaxies.

Caldwell: Patrick Moore's favorite objects not covered by Messier, includes 35 galaxies.

NGC (New General Catalogue): Dreyer's 1887 Catalogue updated to 2000.0 coordinates, includes more than 6,000 galaxies.

IC (Index Catalogue): ICs 1 and 2 were extensions to the NGC in 1895 and 1907 and include more than 3,000 galaxies.

PGC (Principal Galaxies Catalogue): Almost 19,000 galaxies listed brighter than mag 16.

MCG (Morphological Catalogue of Galaxies): Almost 13,000 galaxies listed brighter than mag 16 out of a total of almost 29,000.

UGC (Uppsala General Catalogue): Almost 8,000 galaxies listed brighter than mag 16.

CGCG or Zwicky: Catalogue of galaxies and clusters of galaxies, compiled by Fritz Zwicky; 9,134 objects.

Markarian: 1469 galaxies with an unusually high blue or UV color excess.

Arp: An atlas of 338 peculiar or interacting galaxies compiled by the infamous Halton Arp.

Hickson: A list of 100 compact galaxy groups.

Abell Cluster: A catalogue of 4,073 galaxy clusters compiled by George Abell. Roughly 30 of these clusters are within visual range of amateurs with large telescopes.

The Atlas of Compact Galaxy Trios by Miles Paul may be useful to Supernova patrollers hoping to bag three galaxies at once on their CCD chips. Paul listed 118 objects in his catalogue (see http://www.angelfire.com/id/jsredshift/triplets.htm).

Index

Printed in the United States of America